MATHEMATICS CURRICULUM AND TEACHING PROGRAM

Chance and Data

INVESTIGATIONS

VOLUME 2

Charles Lovitt and Ian Lowe

Curriculum
CORPORATION

Published by Curriculum Corporation
St Nicholas Place
141 Rathdowne Street
Carlton Vic 3053
Australia

Telephone: (03) 639 0699
Facsimile: (03) 639 1616

Reprinted 1994
ACN 007 342 421

National Library of Australia

Cataloguing-in-publication data

Lovitt, Charles.

Chance and data investigations.

ISBN 1 86366 149 2 (set).

ISBN 1 86366 137 9 (v. 1).

ISBN 1 86366 162 X (v. 2).

1. Mathematics — Study and teaching (Primary). 2. Mathematics — Study and teaching (Secondary). I. Lowe, Ian. II. Curriculum Corporation (Australia). III. Mathematics Curriculum and Teaching Program (Australia). IV. Title.

510.7

Edited by Rhonda Idczak and Lynette Wilcock

Design and layout by Rhonda Idczak

Illustrations by David Pearson

Photography by Carolyn Lewens

Cover design by Tom Kurema

Cover photograph by Carolyn Lewens

Printed in Australia by Impact Printing

Foreword

This book and its companion titles, *Investigations: Volume 1* and *Exploring Real Data*, continue the successful Mathematics Curriculum and Teaching Program (MCTP).

MCTP has evolved as a truly collaborative venture, involving input from State and Territory schools and systems, non-government schools and tertiary institutions.

These Chance and Data materials have been developed as a response to *A National Statement on Mathematics for Australian Schools*. In this document, which provides a framework for syllabuses and school programs development, emphasis is placed on the content area of Chance and Data.

The investigations presented in these materials convey problem-solving approaches. They also capture the 'wisdom of practice' of teachers as they seek to integrate new ideas and content into their classroom program and teaching repertoire.

The approach taken also reflects the increasing classroom practice of working towards outcomes. Hence, the materials are very compatible with teaching and learning activities that enable achievement to be assessed and reported, and mathematical competencies identified and developed.

As well as being directly relevant to classroom use, these Chance and Data materials, along with all other MCTP titles, have very strong application in teacher professional development.

David Francis
Executive Director
Curriculum Corporation

Acknowledgements

National Advisory Committee:

Jeff Baxter, Judy Bull, Anne Eyland, Sue Ferguson, Neville Grace, John Hogan, Jayne Johnston, Will Morony, Bruce Mowbray, Howard Reeves, John Rochester, Phil Rofe, Ken Sharpe, Kay Stacey, Max Stephens, Sue Willis.

The project and advisory team wishes to acknowledge that the ideas and materials in these volumes arose in a truly collaborative manner — as a result of very many inputs from very many sources. The following list includes some of those who made major contributions:

Ted Beck, Alistair Carr, Peter Cribb, Murray Cropley, Doug Clarke, Toni Downes, Joy Edmunds, Mark Febbraio, Gail Fitzsimons, Steve Flavel, Sandra Iuliano, Dick Jensen, Mary Keary, Suong Le, Christine Lenghaus, Kay Lipson, Cathy Lovitt, Neil McLean, Pam Montgomery, Huynh Nguyen, Bruce Nulty, Fay O'Connor, Robert Peard, Judith Rivers, Steven Skok, Andrew Tomazos, Diane Tomazos, Els van den Berghe, Sam Watkins, Peter Watson, Bill Wain, Ron Wilks.

Organisations:

The project and advisory team would like to thank the following organisations for their support and for supplying and granting permission to reproduce copyright material.

Angle Park Computing Centre, Australian Bureau of Statistics, Bureau of Meteorology, Channel 9 for meteorology video segments, Channel 7 Sports Department for Australian Open Tennis video segments, Footcams (Footscray City Secondary College media centre), Herald and Weekly Times Newspapers (photograph by Ian Stevenson, p. 254), Mirror Newspapers, Northern Suburbs Dog Obedience Club, St Monica's School, Footscray, USA TODAY Newspaper, Victorian Racing Club.

Anti-Cancer Council of Victoria for the use of the title *Sunsmart?* (p. 342), photograph (p. 342), Sid Seagull illustrations (p. 344 and p. 350), ultraviolet radiation graph (p. 350) © Anti-Cancer Council of Victoria.

Photographs:

Billanook College, Bimbadeen Heights Primary School, Brunswick South West Primary School, Collingwood College, Holy Spirit School, St Raphael's School, Ivanhoe Grammar School, Warrandyte Primary School, Melbourne Girls Grammar School.

Links to other national initiatives

While the material in this package links primarily to *A National Statement on Mathematics for Australian Schools* with its increased emphasis in the area of Chance and Data (the strands of the National Statement are detailed on page vi), the materials also reflect relevant aspects of the developing National Profile on Mathematics and emerging key competencies.

The Finn and Mayer Committees identified key competencies for effective participation in work and work organisation, which include:

- collecting, analysing and organising information;
- communicating ideas and information;
- planning and organising activities;
- working with others in teams;
- using mathematical ideas and techniques;
- using technology.

All these key competencies are reflected strongly in these materials.

An integral part of national curriculum activities is the development of profiles of outcomes in each learning area. Profiles are arranged in eight levels across the years of schooling. There is a general description of each level and a series of more specific statements of expected learning outcomes for each strand or component in the learning area within the national mathematics profiles document.

Teachers should be aware of the emergent trend towards the identification of specific outcomes by which an individual student's performance can be monitored and compared.

Contents

Rationale and overview

The emergence of Chance and Data as major content areas is relatively recent. In particular, there is not a strong history of treatment in primary schools of these topics. Along with greater emphasis being given to the process strands, *A National Statement on Mathematics for Australian Schools* strongly promotes the relevance of these two areas.

As teachers have been exploring these new directions they have been adopting new models of treatment, finding appropriate resources and developing teaching strategies. Among teachers there is already a vast body of knowledge and experience. This program set out to tap into this field-based knowledge, to capture it and to find ways of sharing it for the benefit of all.

Our chosen method of presentation has been to capture and document classroom images for other teachers to consider. It is these that are presented in *Investigations* Volumes 1 and 2 and the partner book, *Exploring Real Data.*

Section A below describes the general characteristics of the package and Section B focuses on its underlying professional development purpose.

Section A: Characteristics of the collection

In analysing what teachers have been trying to do in their exploration of chance and data along with the process strands, we discovered the following common threads:

1 **Learning through problem solving as a method of inquiry**

2 **A holistic approach**

3 **An emphasis on modelling, and simulation in particular**

4 **A high technological profile**

5 **Development of major content ideas**

6 **Treating chance and data concurrently**

7 **Significant integration with other topics and social contexts**

1 Learning through problem solving as a method of inquiry

> [Mathematical inquiry] addresses communication skills, ways of thinking and habits of thought which are explicitly, although not exclusively, mathematical. The[se] mathematical processes... should pervade the whole curriculum, and students should come to realise that it is through these processes that mathematics is generated.
> *A National Statement on Mathematics for Australian Schools,* p. 37

The most essential feature of problem solving is that the solver is doing original thinking, not following previously well-beaten tracks. In this process, learners have to call on all their relevant skills and strategies to find their own path from problem to solution.

For example *Duelling dice* (vol. 2 ch. 8) is presented in a problem-solving manner. In deciding which die is the best to use in a game, students collect data, seek and try to explain patterns, and make and test predictions. This investigative approach is in direct contrast to the simple application of rehearsed probability algorithms.

Above all, 'learning through problem solving' places the responsibility for the learning squarely on the student, and the responsibility for providing the environment and the challenges on the teacher.

2 A holistic approach

It may be unfair criticism, but the typical textbook seems to epitomise the method of dividing content into smaller and smaller chunks. The assumption is that by teaching each atomised chunk separately, somehow it all magically fits together and makes sense in the pupil's head. The types of investigations developed by teachers and collected here do not follow this approach. The titles are not content driven — in most cases some context is explored with a holistic treatment which may cover many interrelated topics of mathematics.

For example, *Population centre of Australia* (vol. 2 ch. 6) covers measurement and data-handling techniques, links to history and current social issues, has very strong links to geography, and, in addition, is presented in a problem-solving context. All these issues seem to be in balance and are treated concurrently. Such is not the case in the type of textbook treatment which isolates and drills a particular mathematics algorithm. Teachers who use the package are invited to debate the relative merits of these two approaches.

3 An emphasis on modelling, and simulation in particular

There are three different kinds of modelling activities within this collection, depending on the purpose of the investigation:

- Simulation models are used to represent a particular mathematical aspect of the situation by a device with the same property, such as the random choice of 12 birth months by selecting from a shuffled pack of cards without kings. Usually the outcomes will be a frequency distribution which displays a lot of useful information, such as spread and average.

The design of a simulation is a sophisticated skill; use of a simulation exposes the student to the nature of the raw data — its range and distribution as well as its mean. The use of a simulation motivates the study of the theory, and provides a check on the theoretical (expected) results.

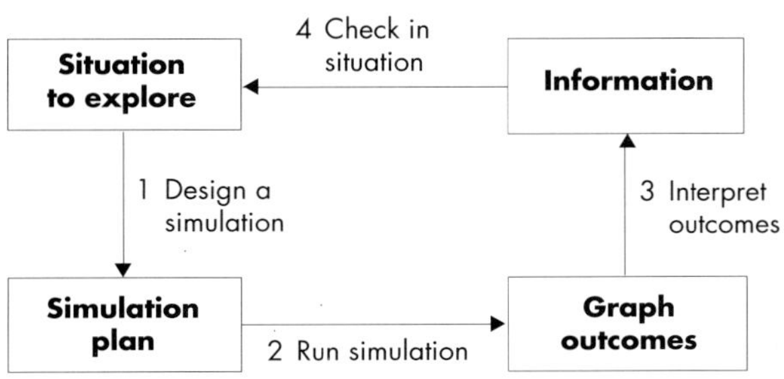

The emphasis on simulation can be seen in *Birth month paradox* (vol. 1 ch. 4) where the agreed answer, after much collection and simulation might be: 'The probability is 0.63 that in a group of five people at least two will share the same birth month'. It is a question for debate as to whether this answer is acceptable. The analytical approach would say that we need the theoretical answer, which turns out to be 0.627. At present nearly all mathematics in schools is an application of the analytical approach. The investigations in *Chance and Data* pose the question, 'Is there not a legitimacy to empirical approaches?'. Can we accept the answer to the birth month problem to be 0.63, knowing that it is not perfect, but certainly close enough for most purposes?

- Descriptive models, such as averages and graphs, tell us about the data we have. They help us to understand it; sometimes they oversimplify it and reduce our knowledge of the actual data. For example, we survey the children in Year 3, and draw a frequency graph of the number of births in each month.

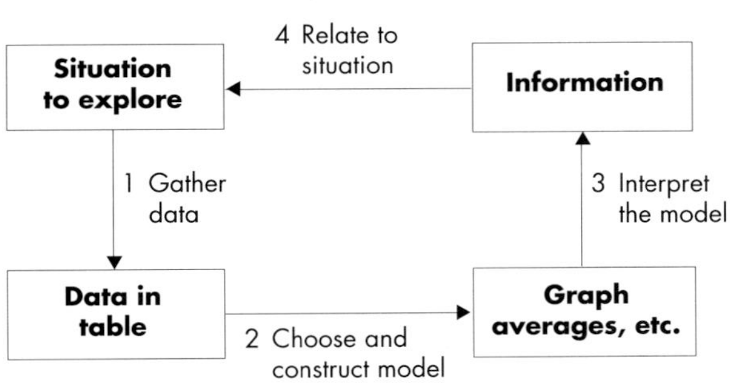

- Analytical models are used when the situation is clearly one which fits a standard model, and for which the appropriate formulas (mean, standard deviation, correlation, binomial distribution…) are both known and understood. The outcomes are usually numbers, and specialised knowledge is needed to interpret them. In contrast to the development of a simulation or the gathering of data, this process requires selection from a known 'tool kit' of models. In the classroom, teachers have found it useful to present these analytical questions as problem-solving tasks. In this way students will be encouraged to draw upon and develop their present understanding.

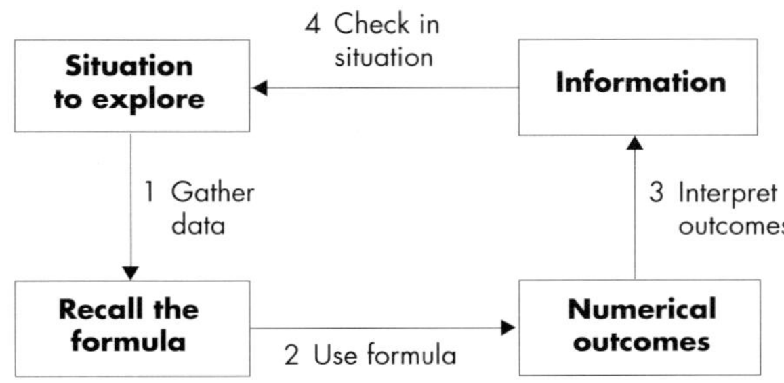

4 A high technological profile

Teachers have consistently sought to include a technological component. The ways they have done this are expressed on page ix. The great variety of methods collectively adds up to an overview of possible contributions from technology to the quality of a learning environment.

5 Development of major content ideas

Teachers have sought to significantly address the conceptual understanding of key mathematical ideas. For example *Horse races* (vol. 1 ch. 3) seriously addresses the concept of natural variability. *What's in the bag?* (vol. 2 ch. 8) focuses on concepts of statistical sampling and statistical inference.

6 Treating chance and data concurrently

Many times throughout the collection, the design of an investigation includes both probability and statistics ideas within the one activity. We believe this integration assists the understanding of both these major threads; they are historically and conceptually linked. For example, *Beetle game* (vol. 1 ch. 4) starts out strongly focused on probability, but the collected data leads into an analysis of stem-and-leaf plots. *Swapping schools* (vol. 1 ch. 3) starts with data collection to answer a question but moves into a probability simulation to help explain why things are as they are. (The analytical response involves exponential decay!)

7 Significant integration with other topics and social contexts

Breaking down the isolation of mathematics is a strong feature of the collection. *Myths and misconceptions*, *Swapping schools*, *Sunsmart?*, *Weather patterns* and *Colas* all start from investigating community issues. *Performance data*, *Population centre of Australia*, *Walk the plank* and *How long is a rod?* all involve other topics of mathematics besides chance and data.

Section B: Professional development purpose

The materials and processes of this package have evolved to support the professional development and growth of teachers. Although it is true that the investigations and data problems can be used directly in classrooms as part of a learning program, this is not their main purpose.

We hope that one way of learning from these images is to study them for what they illustrate about the underpinning value system and beliefs about teaching and learning. To assist this process, we suggest looking at the collection through a series of 'lenses':

- National policy lens
- Technology lens
- Content lens
- Pedagogical lens
- Course development lens

1 National Policy — linking to the National Statement

A National Statement on Mathematics for Australian Schools details the common belief structures shared between the Commonwealth States and Territories around which schools and systems might build their curriculum. The following extract highlights the eight main strands, three process and five content, around which the document is structured.

Attitudes and appreciations
Attitudes deals with the development of positive attitudes towards mathematics and students' involvement in it.
Appreciations deals with the development of an appreciation of the nature, power and scope of mathematical activity.

Mathematical inquiry
Mathematical expression deals with interpreting and conveying mathematical ideas.
Order and arrangement deals with observing and generalising patterns and relations.
Justification deals with explaining and justifying conclusions leading to the notion of proof.
Problem-solving strategies deals with a range of strategies for problem posing and solving.

Choosing and using mathematics
Applying mathematics deals with choosing and using standard mathematical techniques in situations in which mathematics may be useful.
Mathematical modelling deals with the more general processes by which 'real world' phenomena are represented in order that mathematics may be applied to them.

Space
Shape and structure deals with the properties of two- and three-dimensional objects and the relationships between shape, structure and function.
Transformation and symmetry deals with the mathematical equivalent of changes of position, orientation, size and shape, and with symmetries in shapes and arrangements.
Location and arrangement deals with the representation of position and arrangement, including the use of coordinates.

Number
Number and numeration deals with concepts of number and ways we write them.
Computation and estimation deals with the operations of addition, subtraction, multiplication and division and their application.

Measurement
Measurement and estimation deals with the comparison of qualities of objects, the use of units of measurement, and measuring and estimating skills.
Indirect measurement deals with the use of rates, measurement formulae, scale, angular measure and trigonometric ratios for indirect measurement.
Approximation, change and the calculus deals with rates of change, and infinite and limiting processes.

Chance and data
Chance deals with the concepts of randomness and the use of probability as a measure of how likely it is that a particular event will occur.
Data handling deals with collecting, organising, summarising and presenting data for ease of interpretation and communication.
Statistical inference deals with drawing conclusions and making predictions based on both data and principles of chance.

Algebra
Expressing generalisations deals with algebraic expressions as generalised statements.
Functions deals with the general statement of relationships between quantities and techniques for graphing.
Equations deals with setting up and solving equations and inequalities.

A National Statement on Mathematics for Australian Schools, pp. 27, 28

Naturally all these investigations have a focus on the chance and data content strand, but we believe that you will find links to other content strands as well as a substantial use of the three process strands.

Users of the material are invited to consider how closely the materials in the collection parallel the intent of the National Statement. Is the spirit of the document reflected by the images in the collection? For example, does the investigation *Horse races* expose the concept of randomness in ways you believe reflect the principles expressed in the National Statement? Similarly, does the investigation *Radioactivity* adequately demonstrate the nature of mathematical modelling?

2 Using technology to assist learning

There are two main aspects to the way computer technology is used in this project. One is in program support of the two volumes of *Investigations*, the other in providing additional data files for *Exploring Real Data*. Programs are in five formats:

1 Macintosh using Excel templates (2.2 or later)
2 Macintosh (QuickBASIC)
3 MS DOS machines (TurboPascal)
4 BBC (BASIC)
5 Apple IIe (BASIC)

Of the 40 investigations, about 30 have supportive computer programs. In addition, calculators are used where possible. This indicates that teachers have taken many opportunities to exploit the potential of technology.

In *Exploring Real Data* all contexts present data as hard copy, but more extensive data is also provided as a data file on disk. This data can then be manipulated using commercially available database or spreadsheet programs.

The particular ways the computer has been most used are as follows:

- to manipulate data into *graphs* and other displays;
- to *create models* of situations (simulations);
- to *generate* and *analyse data* (either real or from simulations);
- as an *electronic blackboard*;
- to *manipulate parameters*;
- to highlight the *speed* and *efficiency* of computers.

For example, *Walk the plank* uses computers in three distinct ways:

- A graphic presentation allows a demonstration of the game using an *electronic blackboard* effect.

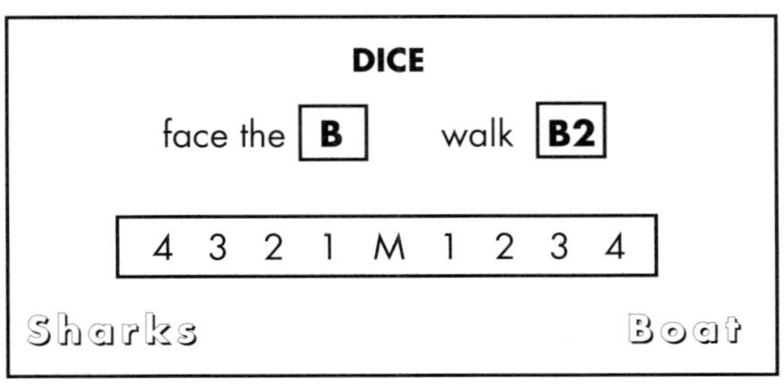

- Once pupils are convinced the *simulation* does exactly what they were doing physically, then the computer can play many games to verify and extend the results the pupils were obtaining. The extra data, *generated* by the program, illustrates the *speed* and *efficiency* of computers. The calculation of a running average and the *graphic presentation* of results also makes analysis easier.

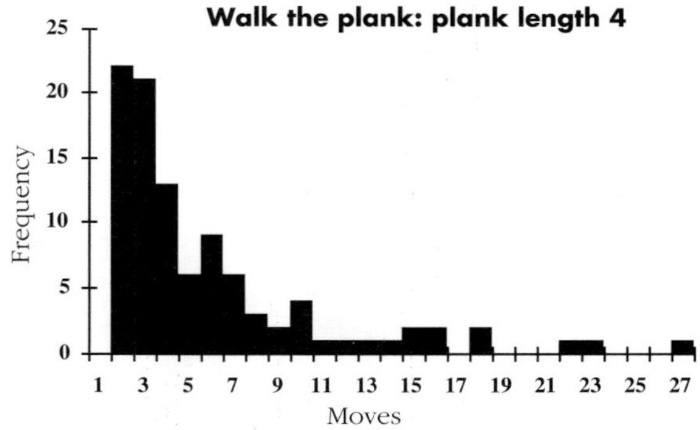

- The program then allows pupils to *manipulate the variable* of the length of the plank and to analyse the effect of this change. In this sense the computer is now a research tool in the hands of the pupils.

Using a computer as part of a simulation investigation

There are three distinct roles for the computer in simulations. These are illustrated for *Beetle game*.

- Demonstration.

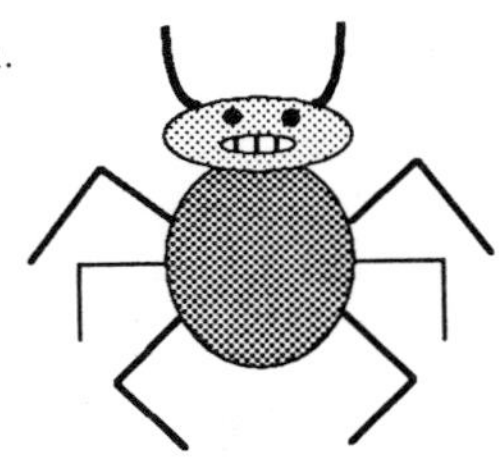

After the children have done some of the work for themselves (for example, they have rolled a die and drawn a frequency graph, or carried out a simulation for a few trials) the teacher shows them in simple slow steps how the computer does exactly the same thing. Large clear graphics are used whenever possible to help convince them of the realism of computer simulation. The screen acts as an electronic blackboard; the display is large, and the demonstration can be repeated or extended in easy steps.

- One trial, seeing all steps take place, but faster.

BEETLE

games to play	50	games	50
fast 1, slow 0	0	die roll	0
body	0	mouth	0
legs	0	feelers	0
head	0	eye	0
parts	0	rolls/game	1
total rolls	2236		
smallest	23	average	45

The step-by-step process in the demonstration is automated and complete runs are made, lasting up to a minute. Now students can also see that the outcomes of each run are summarised in a table, and the table is often the basis of a frequency graph.

- Many trials, very fast, with summary tables and graphs only.

BEETLE

games to play	50	games	50
fast 1, slow 0	0	die roll	0
body	0	mouth	0
legs	0	feelers	0
head	0	eye	0
parts	0	rolls/game	1
total rolls	2236		
smallest	23	average	45

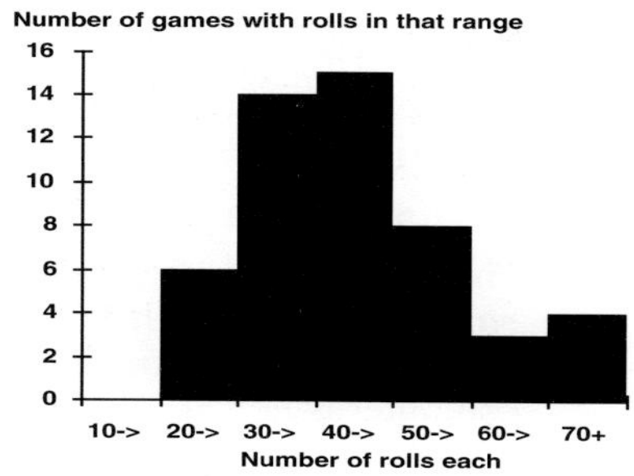

This time the real power of the computer is turned loose on the problem. Many trials can be run at very high speed and the table and graph are redrawn after each trial. In this way the students see the results growing at extraordinary speeds before their eyes. The extremely large number of trials involved means that the simulation results are likely to be very close to the 'theoretical results'. This power enables even very young children to solve quite complex problems through the simulation power of a computer.

3 Mathematical content development lens

Many of the key mathematical concepts of probability and statistics are relatively new to many teachers, hence one purpose of the materials is to allow teachers to develop confidence and understanding in the mathematics. To this extent, most of the investigations have a *Content epilogue* attached to the notes. This section is a little like a mini-tutorial for the teacher, showing how the content ideas develop within the investigation.

The National Statement divides the area into the following three parts, or 'big ideas'. We should firstly be certain about what these mean.

> *Chance* deals with the concepts of randomness and the use of probability as a measure of how likely it is that particular events will occur.
>
> *Data handling* deals with collecting, organising, summarising and representing data for ease of interpretation and communication.
>
> *Statistical inference* deals with drawing conclusions and making predictions based on both data and principles of chance.
>
> *A National Statement on Mathematics for Australian Schools*, p. 27

One way to view the relationship between chance and data handling is the following:

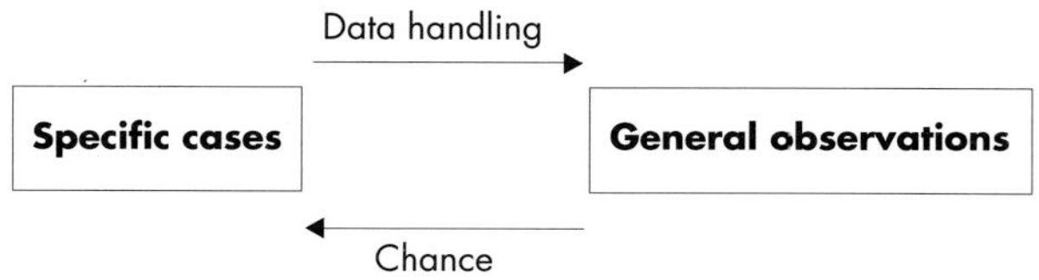

If one child has a birthday this month (a specific case), we may be led to gather data on many children's birth months. This could lead to a general observation that all months would be equally likely. From this general observation, we can predict that the chance that a child was born in June is 1 in 12. So going from the general to the specific involves ideas of chance, but the other way involves techniques of data handling.

Another way to look at the relationship of the three big ideas involves a distinction between two uses of statistics:

- describing a sample (descriptive statistics) and
- making inferences about the general population based on the data from a small sample (inferential statistics).

The first requires graphs, averages, and ways of describing the spread of the data. The second needs ideas of probability because a sample selected from a larger population may not proportionally represent the total population. (For example, the children in the class may not have birth months equally shared over the year, even though the general population does.) Ideas of probability are used to take the chance of a 'biased' sample into account.

Probability and statistics are conceptually very tricky areas for many learners. One reason is the language in which problems and applications usually appear; the terms used to describe situations involving probability and statistics are often not clear to the uninitiated. Students from non-English speaking backgrounds will be at a particular disadvantage when vernacular expressions or complex English constructions are used to describe conditions for a probability problem.

A second reason for difficulty in acquiring the concepts of probability and statistics is that there are many types of displays and summary numbers, several applying to some types of data and not to others. Thirdly, most situations involve uncertainty and hence the data is messy and unpredictable. In addition, attempts are often made to predict the future using the tools of chance and data and information gleaned from the past; although long-term outcomes are stable, short-term outcomes remain unpredictable.

4 A pedagogical lens

Chapter 3 of *A National Statement on Mathematics for Australian Schools* articulates four learning principles; these are the basis of all of the investigations described in this book.

The four learning principles are:

- Learners construct their own meanings from, and for, the ideas, objects and events which they experience.
- Learning happens when existing conceptions are challenged.
- Learning requires action and reflection on the part of the learner.
- Learning involves taking risks.

All the investigations and problems utilise a wide variety of teaching and learning strategies to try to achieve the above. These are summarised at the start of each investigation under the heading *Teaching and learning*. The vocabulary chosen to describe these features includes words and phrases like 'group work', 'physical and active involvement', 'visual representation', 'story-shell', 'firsthand data', 'links to social issues' and 'personal focus'.

The materials make no claim that these represent better quality learning; this decision is wholly left to the user. The invitation, however, is to trial the activities and reflect on the effects in the classroom, before making such decisions and determining whether particular features deliberately added to an activity *do* activate the four principles of learning detailed above.

Reasons why teachers have injected these teaching and learning features into the materials include:

- Students take an *active* role, both mentally and physically. The student *actively builds understanding* by using investigations to explore ideas, to search out the way things work and fit together, and to discover how these new ideas are linked to other ideas already understood.
- Students are *responsible for creating their own meaning*. This cannot be taught or memorised, and hence students must be expected to actively search for meaning, and have the right to expect the classroom to place them in situations where they can continue that search. In such classrooms learning will be fun; curiosity and a wide-spread natural desire to explore, find and explain patterns will develop out of experiences that are rich in opportunities, open-ended and non-threatening.
- Each student's search for understanding develops to the point where the students actively seek to *master a set of mathematical tools* (a tool kit) that can cope with a wide range of real, practical situations. To this should be added an ability to select and adapt a part of the tool kit to suit any problem being studied.
- The learners engage in the activity, preferably *in company with others* with whom they can consult along the way. This opportunity for sharing of goals and outcomes leads to many opportunities for sharing understandings through *peer teaching*.
- The learners are *involved in the choice* of the problem, and in developing their own reasons for wanting to do it. This process of negotiation could be little more than responding to a challenge or interest in a current event, but it is essential for the commitment of the learners to the task.
- The importance of *concept learning* is to the fore in the materials. Concepts develop as they grow in the number of connections they make to other ideas held by the student. Thus, making links to other ideas, mathematical and otherwise, is an important part of the process. This process is a little like forming a mental map of a town into which you have just moved; you gradually see connections and the way the streets and landmarks are related. Understanding takes time, requires a lot of individual effort and sometimes help from others, but most of all requires active seeking by the learner.
- The careful use of language plays an important role in learning. By itself an activity is not learning; it requires *reflection* on the meaning of what has been done. This requires a chance to *discuss* ideas with each other, to hypothesise, to justify, to dispute and to form a consensus about the ideas. In short, it is very important that students have a chance to discuss with each other, and with their teacher, so that they come to understand the ideas involved. This is particularly important for students of non-English-speaking background, since these students have to learn new vocabulary and its meaning as well as working in an unfamiliar language.

In summary, teachers are invited to suspend judgment, trial the approaches and strategies, and then reflect on their suitability for inclusion in their long-term repertoire.

5 A course development lens

One of the thoughts behind the development of these materials was the desire of many teachers to go an extra step beyond the type of lessons in the RIME and MCTP programs. These were designed to demonstrate specific teaching features, but in the main were one-off type lessons, with no suggestions as to how they might be integrated into a unit plan.

Unit planning was not one of the criteria the project sought to address, but there was a strong emphasis towards showing significant treatment of the mathematics content itself, not just the teaching strategies. Unit planning is envisaged as the prerogative of the states and local systems, not a national project, which might incorrectly be seen as trying to establish a de facto national curriculum.

The middle ground exemplified by the investigations in this collection is an extended piece of work that might take one or two weeks, but stops short of being a definitive unit. After you have trialled these extended investigations you can determine if there is a place for them in the course your school is developing, and for the structure to act as a 'template' for investigations of your own design.

Investigation outlines

VOLUME 1

Chapter 1 Intuitive probability

Clothes line

Clipping clothes pegs onto the side of a paper bag provides a tangible, visual and colourful display of possible combinations. Generating these at random engages pupils in intuitive recognition of the probabilities involved.

Colour dice

A story-shell about a lolly machine that puts coloured lollies into bags at random is the context for this investigation. Coloured sticky dots are placed on dice to simulate the selection of lollies. Problems about the likely composition of the bag are posed and modelled.

Crazy animals

A well-known book design which mixes and matches parts of animals to make 'crazy animals' is the stimulus for the dice games, problem solving, graphing, computer and language work which are all woven into this rich and active investigation.

Likely choices

Children draw pegs or counters from a tub and informally discover the chances of particular combinations occurring. The language of chance is a featured part of the activity. Pupils' guesses of what might be drawn out are an excellent indicator for the teacher of their intuitive understandings.

Peg spin

Pegs and coloured disc spinners are the concrete aids that expose a range of probability situations, in particular, the concept of unequal likelihood. Can students predict what will be selected from the colours on the spinner?

Early probability

This is a collection of a range of simple activities, all with a probability component. Each is of short duration, but all are concrete and active. The teacher can select one or two from the collection for each period. They are worth revisiting regularly.

Chapter 2 Probability calculations

Sums and differences

Rolling and finding the sum or the difference of two dice is a well-known probability activity. But the variations of this theme in this activity are quite clever in the challenges posed to pupils. There is much analysis, creative thinking, group work and problem solving as pupils create and play a number of dice games.

Myths and misconceptions

Probability is the source of a rather large range of misconceptions. Many people believe that if you toss four heads in a row, then in the next few tosses there will be more tails — 'to even things out'. A range of such misconceptions is offered to pupils. For each they find out what people believe and then design a simulation to learn the truth.

Probability smorgasbord

Just as a food smorgasbord allows people to choose items of interest and attraction to them, so this smorgasbord offers a wide range of small investigations from which pupils can choose. The items on this menu, however, have all been chosen because they offer a selection in probability skill and concept development.

Chapter 3 Applications of chance

Horse races

The natural variability of dice rolls is the basis for this investigation. The context of a 'race' to see which number will be rolled five times first adds interest. Each number is equally likely to win, but the variability can be very wide; sometimes one number is rolled five times before another has occurred once. Computer extensions and graphing are additional features.

Odds: fair and unfair

Pupils leaving school need to understand concepts like odds and a fair return. A lack of understanding of these concepts could lead to exploitation. This investigation uses a simulation to expose the results of different betting strategies. Building the investigation around a major event such as the Melbourne Cup adds stimulus and a sense of reality.

Swapping schools

How often do students swap schools? One statistic quotes that one in eight Australian pupils changes school every year. Building simulations that reflect this statistic shows the patterns of change over many years. This model data can then be compared with real data gathered by pupils on themselves and their parents' generation. In so doing, pupils gain an appreciation of the increasing mobility of the Australian population.

Radioactivity

The world of nuclear science has a mystique that can be partly removed for pupils by this activity. The concepts of radioactive decay and half-life are cleverly elucidated using a simulation that involves pupils role-playing atoms over time. Physical involvement and group work enhance the learning environment.

Medical testing

Medical testing for a disease (like AIDS) often produces the scary result (for the patient) of a false positive. The probability basis for this circumstance is revealed through a simulation. The underlying mathematics is known as Bayes' theorem.

Chapter 4 Modelling with simulations

Walk the plank

'Victims' walk a plank according to random dice rolls, finish up with 'the sharks' or in the safety of the boat. If the plank is too long the game is boring; too short and it is over too quickly. A probability simulation helps pupils decide on the 'best' length. The story-shell, group work and computer program all help the structure of the investigation.

Beetle game

This simple dice game has been played for generations. Make a beetle by rolling a six for the body, five for the head, and other numbers for legs, feelers, mouth and eyes. The story context of a fund-raising game at the school fair challenges pupils to discover the distribution of likely results and to understand the probabilities involved.

Birth month paradox

Five people meet at random. What is the probability that at least two of them share the same birth month? (This activity is a variation on the famous birthday paradox.) Students nearly always incorrectly estimate the probability as less than 50%. They are challenged to analyse their preconceptions as more and more information is obtained, thus exposing the paradox.

Winning streaks

Two players equal in ability play each other regularly at tennis. One of them suddenly wins five games in a row. Could this player now claim to be the better, or could such a result be expected within the natural variability of chance events? Generating data through simulations allows pupils to analyse the probabilities involved.

Matching

If you had ten name tags for ten photographs and mixed them up, what is the likelihood of matching the correct name with each photograph by chance? Not very likely, but how many of the ten could you expect to get correct? This simulation shows that the expected answer is exactly one. But what happens if you had nine photographs, or three, or perhaps 52? The simulation reveals the unexpected answer.

Chocolate-chip cookies

If you put 20 chocolate chips into a mixture to make ten cookies, the average will be two chips per cookie. But because of variability, it would not be a surprise if some cookies had three and others only one. The underlying challenge in this simulation is finding how many chocolate chips should go into the mixture to ensure at least two in each cookie.

VOLUME 2

Chapter 5 Using personal data

People graphs
Students make physical graphs by grouping themselves using such criteria as how they came to school or in which month they were born. This physical representation gradually gives way to more abstract forms of representation. Secondhand data is also introduced and there is much opportunity for problem posing and problem solving.

Performance data
Statistics are often used to measure human performance. In this investigation, data is collected on pupils' performance in a variety of tasks. This firsthand data is then analysed to indicate if any improvement has occurred. Graphing of errors to show trends is a particular feature.

Soup kitchen
Children use data collection techniques and graphing to arrive at a recipe for soup using only ingredients liked by a majority of the class. The firsthand data and personal involvement are positive features of this statistical investigation.

Chairs and tables: just right?
Do your children wriggle on chairs unsuitable in size, or bang their knees on tables too small for them? Providing furniture to match the varying sizes of pupils throughout a school is a surprisingly mathematical task. Pupils learn and employ a range of statistical methods in exploring this question.

Groups and sorting rules
Physical graphing by putting all pupils in groups is easy for groups that don't overlap, such as boys and girls. But what if one group is boys and the other is all those who ride a bike to school? How the groups can be represented to allow for the overlap is the source of much problem solving.

Bikes, monkey bars and skeletons
The three parts of this investigation have a common theme of body lengths, and how the design of playground equipment takes account of these measures. It employs statistical methods such as simple column graphs and stem-and-leaf plots to help analyse the data collected.

Popcorn
Children decide to make some popcorn to raise money for charity. 'How much should they make?' 'How much should they charge?' These questions are answered using simple but important statistical techniques. Ideas of sampling are to the fore in this investigation.

Chapter 6 Using secondhand data

Video data
Most data is presented to pupils in numerical or graphical form. In this investigation, problems are posed to, or by, the pupils, and they realise that the data they need is contained in video footage of the provided snippets. For example, a video snippet of cars on a freeway can be the data source for many questions such as 'What is the average number of passengers per car?' or 'What is the most popular colour for cars?' Pupils have to collect the data they need by counting, measuring or timing directly from the screen.

Population centre of Australia
A curious newspaper article about shifting population trends is the stimulus for this investigation. Estimating where Australia's current population centre is, and in which direction it is moving, provides the stimulus for analysing supplied historical data. As well as learning some interesting statistics and using computers, pupils also find out a lot about their country and some of the social issues underlying census data.

Top of the class
Statistics have always been used in test score analysis to rank pupils. But there is a range of statistics available, and such statistics do not always produce the same rankings. Indeed this investigation shows how several different techniques applied to the same set of scores can produce quite opposite conclusions. Group debate is also a feature of this activity.

Dying for a smoke
This investigation provides much data on the incidence of a range of diseases, cancers in particular. Graphing and other statistical techniques are employed to allow pupils to see the 'story' that is told by such data.

Sunsmart?
Analysing published data provides a compelling and sobering statistical picture of the incidence of skin cancer, particularly in Australia. Pupils not only practise statistical techniques, but also address the underlying factors behind skin cancer in a way that educates them about methods of reducing the risks.

Chapter 7 Using databases

Database debut

Children build a database containing information about themselves, and based on questions which they themselves design. By entering this data into a computer, the children are encouraged to ask the kind of questions that can most easily be answered by manipulating the database.

Pocket money

An initial small-scale investigation raises a number of questions about the amount of pocket money received by children in the class. On closer study of the data, children are led to ask more questions and see the need to further develop their data collecting and analysis skills.

Weather patterns

One organisation that perhaps collects and interprets more data than any other is the Bureau of Meteorology. This investigation highlights some of this data and allows pupils to pose and answer a range of hypotheses based on patterns they observe in it.

Cemetery maths

The local cemetery contains a vast store of data which allows pupils to pose and answer many questions about population trends and the probabilities of death at different ages in different times. It is a very rich investigation which touches on many issues besides practice of mathematical skills.

Chapter 8 Sampling and inference

What's in the bag?

Sampling techniques are to the fore in this interesting and challenging investigation. How many of the items contained in a bag do you need to see before you are reasonably confident of predicting all the contents? Lots of simulation, estimating, predicting and testing, with computer support, are included.

Duelling dice

The counter-intuitive result is the highlight of this investigation. Firsthand data generated by a series of games uncovers a generally unexpected result. Explaining the result gives much exposure to probability theory, but this is achieved in an informal and enjoyable manner.

How long is a rod?

Before the days of well-defined units of measure, an English unit called the 'rod' was determined by adding the foot-lengths of 16 men. Not surprisingly, the unit varied in length. But how much variation would you expect to occur? This unit of length generates much interesting history as well as mathematical ideas.

Colas

We all suspect sometimes that those who claim to have a favourite brand of some commodity might not be able to tell the difference under test conditions. Yet advertising for products is often premised on the assumption that people can distinguish. In this investigation, the proposition that pupils can tell the various colas apart by taste is put to the test. Ideas of confidence limits and sampling techniques emerge as being important here.

Major mathematical content and recommended year levels

VOLUME 1

VOLUME 2

5 Using personal data

These investigations use information obtained by children themselves, often about themselves, as the data for statistical treatment. To this extent only they differ from the investigations in chapter 6, where data is provided for interpretation. (Chapter 7 offers experience in both kinds, through the medium of the computer database.)

Groups and sorting rules provides one way to help young students start to use and understand the ideas of simple classification, which is a fundamental idea in the process of making sense of data. The other investigations use data gathered by students in response to questions posed to them or by them. The teachers who developed the ideas went in these directions because their students led them; feel free to respond in your own way to the real interests of your students.

The sophistication of the statistical process grows from the simple asking of one question and moving people around to display the results (*People graphs*) to the design of a questionnaire to find what type of *Popcorn* a whole school prefers, and how much they would pay for some amount of it.

Because some of these are ideas based on real projects tackled by the students (e.g. making and selling popcorn, making a popular soup), the surveys and their analyses are part of integrated themes, involving much measurement and calculation, language, etc. Again, you could use these treatments here as models for your own development of a theme, with surveys, that 'belongs' to your own class.

From *A National Statement on Mathematics for Australian Schools*

(Overview) Most people need only occasionally to collect and handle raw statistical data and convey such data to others. Nonetheless, students would have considerable opportunity to collect, organise, represent and interpret data from situations of interest to themselves, in order to answer questions which they and others have posed. This should help them to see a range of purposes for collecting data and to develop an understanding of data collection and handling processes.

(Band A) Initially there should be an emphasis on the use of the actual objects or physical representations of objects or measurements, although children should gradually move to simple pictorial, block and bar graphs...

(Band C) In conducting simple surveys, the processes involved in gathering the information are as important as the results obtained. Children should decide:

- the type of information that is to be collected, and the people (things) to be surveyed;
- how to collect, record and organise the information;
- an appropriate way to display the information to others: for example, a table, diagram or other representation.

Children should learn that data can be displayed in a variety of ways and that the choice of display depends upon the question being asked of the data. Graphs should not be regarded as an end in themselves; rather they should serve purposes which are clear to children. (pp. 164, 165, 168)

People graphs

Years 1 to 3

How can you introduce graphs to very young children in a meaningful way? This investigation offers a practical and very personal meaning, first using the children themselves as elements of graphs, and then using sheets with their own names as elements. The two situations suggested — birthdays (graphed by months) and transport (ways of getting to school) — are models for many others.

Features

Teaching and learning

- Personal and secondhand data
- Whole class physical activity
- Problem solving
- Problem posing by students
- Representing data on cards
- Use of computer or fax

Major mathematical content

- Simple graphing representations
- Learning to model data
- Types of graphs
- Interpreting graphs

Mathematical content is further discussed in the *Content epilogue*.

Equipment and preparation

- Sheets of plain paper (either A4 or half A4)
- Felt pens
- If possible a space on a wall to display these sheets
- Small bits of Blu-Tack ready for students to use to hold the paper in place
- If you have access to a fax machine, arrange with a teacher in another school to send you data

Expected timing

- Three or four sessions, even up to seven using further situations to graph

Learn to use the *Chance and data* software disk if possible, as described in stage 10, so that this tool can enrich students' learning.

A teacher's story

This starts with very concrete graphs and moves steadily towards more abstract representations, using the power of technology as a support. I wanted to get across an initial feeling for graphs and how they store information. I used *firsthand* (personal) *data* and represented them by a physical group situation. The people graphs are very real for students. I tried two different contexts, though there are many possibilities. Once the graphs were created, I asked lots of simple questions which could be answered from the graphs — many questions were generated by the students.

Another development was to use *secondhand data* from another source. We started with the class next door, then used a fax link to another class, and went on to use the data stored on the software disk. This process extended the principles of graphs and problem posing and solving. The children took one step back from being so personally involved, focusing more on the mathematics. They also learned a lot about the transmission of data using modern electronic technology.

Finally, the students wrote stories as a summary of what they had learned. I used these to gain an insight into their understanding and to help plan where to go next.

Comments

I gained experience in teaching a variety of mathematical concepts using the children as a starting point and developing the concepts over a series of lessons.

The lessons showed how successful problem solving can be. An innovative, active, exciting lesson makes the work great fun and the children are willing to participate and try their best.

I gained the ability to adapt ideas and activities to my own situation. I also developed the skills to use materials to highlight mathematical concepts.

The richness of this lesson is a result of how it is done. Whether the students are measuring feet, counting pets, listing birthdays or whatever, the activities are the essential ingredient. The way each part of the lesson is conducted is the key.

The lessons flowed nicely from one step to the other without either the slower student falling behind or the smarter student flying ahead.

Active participation allows students to engage in learning by discovery.

Summary of the stages in this investigation

1 **Choosing a context: birthdays**

2 **Sorting into groups**

3 **Wall or floor graphs**

4 **Manipulating and interpreting the graphs**

5 **Another context: coming to school**

6 **Graphs with people**

7 **Making a record of the information**

8 **Extending the database to other contexts**

9 **Secondhand data: the class next door**

10 **Secondhand data: technology**

11 **Writing stories**

1 Choosing a context: birthdays

I started with the children's birthdays as this data is simple and easy to collect. How they come to school or the number of pets in a household are also good. What you need is easily obtained information that creates about five to 12 categories.

I want you all to write your name and birthday on your own sheet of paper.

Students are given a piece of paper (A4 works well).

Who has a birthday in January? Who has a birthday this month?

Felt pens are very useful for visibility, especially later when the graphs go on the wall.

2 Sorting into groups

Find all the others in the class who have birthdays in the same month as you.

I used an outdoor area near to our room.

I let students discover methods for classifying and organising themselves. This involved them physically moving about.

If students have their pieces of paper with their name and birthday on them, this becomes a little less chaotic. In addition, the paper is an abstraction of the data, which helps the transition into graphs.

After groups had formed, generally in clusters all over the place, I asked each group to announce their month and how many birthdays.

Children managed to group themselves, but found it difficult to view other groups and get an overall view. I was able to use this fact to interest them in a wall display.

Are there any months not represented?

Which is the largest group?

Which is the smallest?

This first clustering is a form of graph. But so that we could clearly see all the information, I challenged the groups to reorganise.

3 Wall or floor graphs

Can you now pin up your sheets in a useful way? We want to be able to see the pattern clearly.

The children organised the cards and left a space for the months in which there were no birthdays.

I provided little blobs of Blu-Tack to hold the sheets in place.

Students expect, perhaps intuitively, that the data will be evenly spread. However, it is not unusual to find a month with four or five birthdays and others with none.

Students often create an informal bar graph, nearly always left to right, and often with the current month on top (not January as I would have initially expected).

4 Manipulating and interpreting the graphs

This wall graph provides our first 'database' which can now be used in a variety of problems to highlight the different ways in which data can be presented and used.

Each new arrangement involved all students coming to the wall to re-position their cards.

Can you organise your sheets in a different way?

Can we change an across pattern into an up and down pattern?

Which month is the most common?

Some students want to put the most frequent on top, some want to use the order of months, and some want to put the birthdays in the order of the dates within each month.

Finally I moved them to another stage in symbolising the data. I asked them to draw a diagram in their books.

I have realised the importance of letting the students hypothesise about what they think is being shown, and what the results will be, to allow them to draw more relevance from the lesson.

5 Another context: coming to school

It was really sunny this morning so I walked to school. Usually I come in my car. How do you usually come to school?

I listed the suggested methods of transport on the board. I then handed to each student a blank sheet of paper.

Write your name on the paper, and how you usually come to school.

A4 sheets cut in half are a good size to use.

Some children may use more than one mode of transport each day. To handle this, students may suggest a group for the combination, for example 'bus and car'.

6 Graphs with people

How could we sort ourselves into groups which show how we come to school?

I listed the suggestions, and we tried several ways of grouping them. We discussed each approach as it was tried, and modified the list as a result.

Each group had the problem of making sure that everyone using that particular mode of transport was in their group.

Real problem-solving situations will arise if some children do use more than one mode of transport to get to school.

Now that you have made your groups, what can you say about how children in this class come to school?

Are there things we could do to make it easier to comment about our groups?

There was quite a bit of confusion at first when my class tried to sort themselves out, but eventually they were able to successfully organise themselves into groups. Having the information clearly written on paper helped enormously.

The presentation got the children excited and enthusiastic; as a result they produced wonderful work.

7 Making a record of the information

We need to save the information about how we come to school in case we want to use it again or show it to another class.

In your groups, discuss possible ways to do this. Let's try some suggestions. Which will we start with?

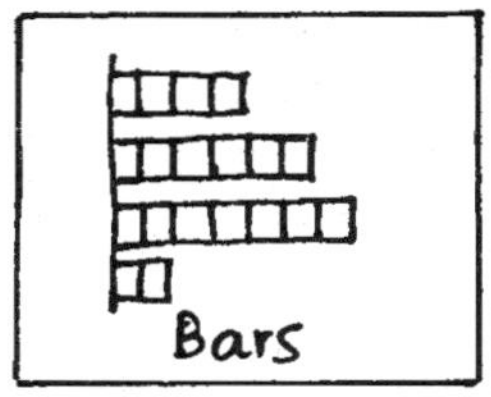

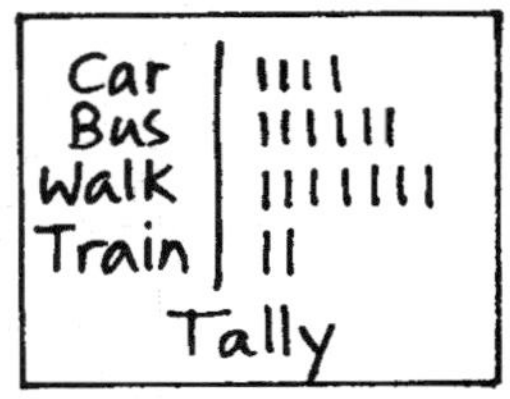

I spent time discussing ways of saving information, encouraging suggestions from the students. I listed these on the board as I went along.

Be prepared to try some of the suggestions. It may pay to anticipate what might come up and have likely materials on hand. I often find that recycled materials from the art room are very useful.

What can you say about how the class usually come to school?

I was pleased to see the children naturally used their group skills during this discussion, quickly nominating a leader, recorder, reporter, and so on.

When students' ideas are ignored or rejected they soon understand that their job is to guess the 'answer' you had in mind.

8 Extending the database to other contexts

The database can be used over an extended period by adding new information, or other data can be collected and manipulated in similar ways.

Before moving on to secondhand data, I repeated the exercise with:

- number of different pets
- family size

Who is in a family can raise issues of definition, and possibly some sensitive areas. For example, an older sibling who has left home, or granny in the flat, or...

In a traditional textbook, instead of the children discovering the different types of graphs, they would be told what the graphs were, and then practise constructing them.

9 Secondhand data: the class next door

The first level of access to secondhand data was by producing a list of suggestions for the 'class next door'.

This apparently small step is important, but I found that the students easily accepted the reality of secondhand data based on the experiences they had just gone through.

Do you think that a similar graph of birthdays made by the class next door would look like ours?

What would we have to do to find out?

Let's make a plan. What will we do first? After that?

How could we do that? Who will do it?

And the next step?

Children found this activity very interesting and realised that each class's graph would be different.

I involved students directly in the process of the investigation by getting opinions. I asked them to explain why they thought the situation would be as they suggested.

I listed the suggestions on the board, adding any obvious omissions.

Several presentations of the data are possible, but back-to-back bar graphs work very well.

My children suggested a plan which I thought would be quite chaotic and have little likelihood of success. To my surprise, when they put the plan into action, they knew just what they wanted to do and the outcomes were most impressive.

Since wall graphs take up so much space, the suggestion is usually made that for easy comparison we use smaller pieces of paper.

Now you have shown both sets of information as graphs. How is that class different to us? How are they the same?

Students commented and wrote a summary of any differences they noted. I kept emphasising that it was the graph that helped us notice the differences.

10 Secondhand data: technology

An important aspect of data handling in our information-rich society is the way technology is used. For this reason I made great efforts to include technology — although all the mathematical messages about graphs can be achieved in other ways.

Fax machine

I had organised to ring another teacher far away in another state. We looked the place up on the map. Then we rang the teacher and asked his class to collect their data and fax it to us.

How is that class different to us?

How are they the same?

Most students were amazed at the fax: 'How did the pictures go through the lines?'. They were all keen to have a go!

Computer database

The available chance and data software disk contains a class list contributed by a school. I found it very valuable for the students to select the data from the computer menu, print them out and compare their class with this class.

How many are in the class? What information does the data give us about them?

	NAME	B'DAY	TRAVEL	
1				
2				
3				
4				
5				
6				

I transferred the data file into the database program (MS Works) which runs on my school's system.

The power of a computer program to present the data in various forms was a feature I wanted to promote.

Let's see how many children have birthdays this month.

The students greatly enjoyed the option of putting their data into the machine and printing them out. They were then able to graph the data in several different ways.

Computer graphing

The software program *People graphs*, if available, is a simple graphic program which builds a column graph of birthdays, by months, for your class.

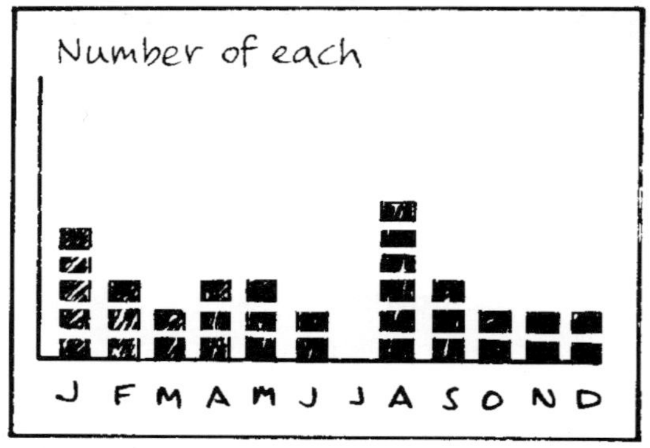

The students easily accepted that the computer was simply doing the same thing they had done!

11 Writing stories

I asked students to write a story to an imaginary friend in another school saying what they learned and what they liked the most. If you used the fax connection with another school, then this is an ideal audience to write to and then actually fax the letters. Maybe this could lead to an interchange of letters between classes.

This was a wonderful chance to link creative writing to mathematics. We were learning to use computers for word processing so I took the opportunity to do it that way.

I found that having a real audience for their writings was more powerful than just making notes.

Children like to have some control of planning these hands-on activities; such activities can be made quite worthwhile over all levels of schooling.

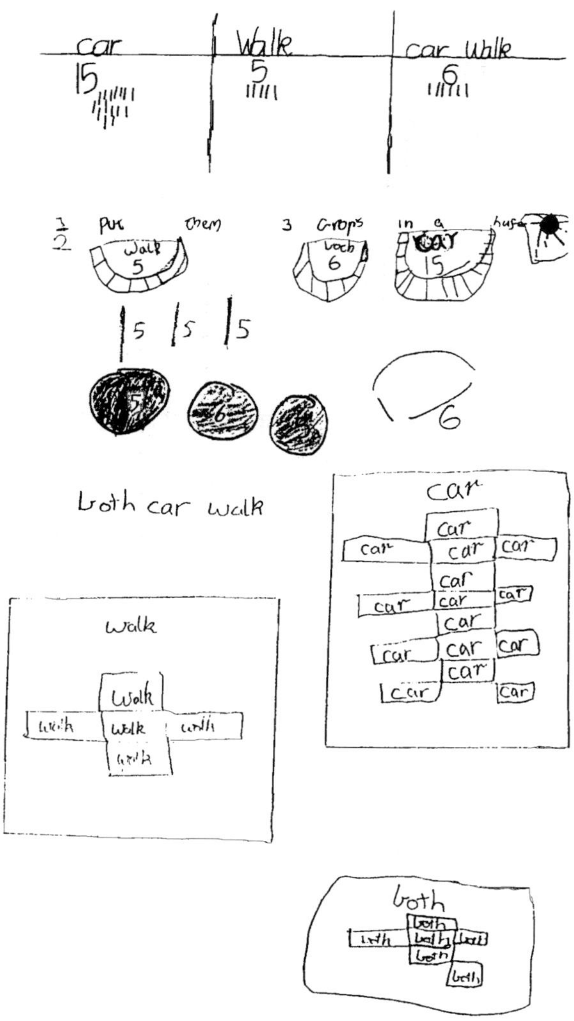

Now I know that when you do a graph you do it claerly so that when you put it up on the wall you can see it and you use textas too. We went outside and put ourselves in order because we were talking about how we get to school. There was Car, Walk and Both. We got together with Mrs. Robertson's grade and did a big graph with our groups about Car, Walk and Both.
....Emily 1G

Content epilogue: background information for teachers

Frequency graphs

This set of investigations aims to introduce two concepts to young children:

- the representation of data in the form of simple statistical graphs;
- the meanings of statistical graphs in terms of the data they represent.

These are both very sophisticated ideas, and will be developed much further in later years. In particular the notion of *frequency* (how many times something has happened) is central to what is being graphed.

In a simple column graph, each column shows data for one person. For example, it may give a column showing the height for each student. The labels along the horizontal axis are names, and the vertical scale gives centimetres. This is quite different from the type of graph used in this sequence of lessons.

Column graph

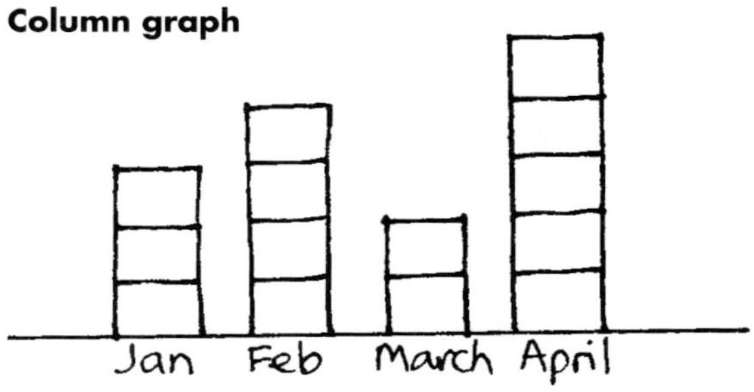

The column graph we are using shows the frequency on the vertical axis. Each child is part of the column, so that the column height shows the number of children who fit into that category (for example, have birthdays in that month).

The graph gives an overall picture of some characteristic of the whole group. Learning what information can be read from the graph, and what cannot, is a valuable part of learning to read (or interpret) a graph.

Natural variability

Children tend to have expectations that births are equally spread across all months — in a typical class this is most unlikely. It is much more likely that a great variation will exist in the number of students born in each month. This 'natural variability' is a common feature of most real life situations involving chance. (For example, if you roll a die 30 times you are very unlikely to get each number exactly five times!)

Sets

Every child has one, and only one, birth month. However, ways of coming to school or types of pets they own may overlap (for example, walk and car or cat and dog). This situation is often shown as an overlap using a Venn diagram.

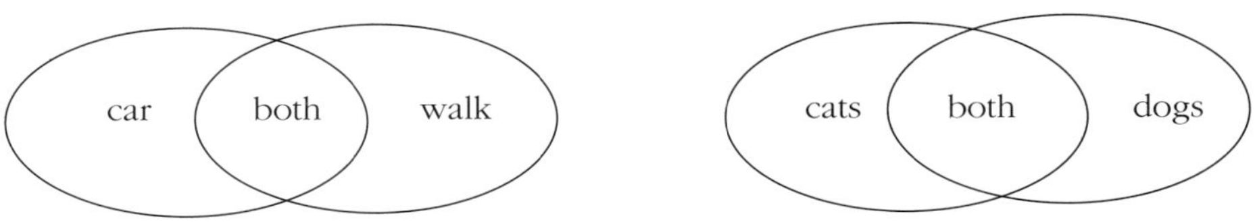

The students who own *both* cat *and* dog are in the overlap, called the 'intersection'. Those who own *either* a cat *or* a dog (*or both*) are in one circle or the other (or the overlap), called the 'union'. This idea of overlapping sets is a feature of the investigation *Groups and sorting rules* on page 264.

Performance data

Years 2 to 7

How can students use graphs and statistics to show whether or not there has been an improvement in performance? These activities provide models of the use of graphs that can be used to display performance in many areas. In drawing and interpreting the graphs, students solve problems and give meaning to useful mathematical models.

Features

Teaching and learning

- Connecting graphing with other learning areas
- Estimating
- Personal firsthand data
- Studying the effects of feedback and group pressure
- Using a modem or fax to get data
- Group problem solving

Major mathematical content

- Using technology (calculators and computers)
- Refining and clarifying questions
- Measuring time and distance
- Intuitive concept of speed
- Tabulating data and finding averages, ranges, etc.
- Choosing the best type of graph
- Graphing and interpreting, looking for patterns and trends
- Conducting mathematical arguments
- Writing reports

Expected timing

- The first activity takes at least a week; the others take about three sessions each.

Comments

The content was real to the students. It was easily understood by all ages. There was a wide range of mathematics for the various ages to tap into.

None of the older children had a deep understanding of 'average' prior to this study. It was necessary to explore this concept as we proceeded and the children seemed to find it very easy to understand using their personal data. They could suddenly apply ideas of average to other settings.

It was wonderful to see an activity which generated sustained interest and determination, and which developed such valuable skills. It re-emphasised the value of group work and outdoor activities.

They were using some quite difficult skills of problem solving, and creative thinking in terms of graphs and use of data. It showed how graphing could be linked to many investigations.

A teacher's story

My major motivation was to link statistics, other mathematics topics and assessment information. The students and I were convinced of the value of these connections by using activities like the following.

Break your own record

The children chose their own physical task to measure their performance, hoping to see some improvement in a week. They tabulated and graphed their scores on ten trials per day, and worked out ranges and averages of different kinds.

These children were captivated with the activity. If we didn't continue during the morning, they would take their books out and use the lunch break to collect data. If they had any free time, they chose to work on their mathematics. It wasn't just the physical activity which inspired them; they were equally enthused about totalling their scores, looking at patterns of improvement, or graphing results.

The 40-second walk

I like this because it is active, the data is firsthand and personal, it involves groups, and it includes going outdoors. The activity is oriented towards concepts rather than skills, so no formula is developed and students operate informally and intuitively. In this way it is a good basis for later development of the formula or rule for speed measurement. A major strength is the non-threatening structure which allows all students to become involved. Another strength is that students' improvements are obvious and reinforce for them a sense of accomplishment.

The time for each 'walk' includes an error. If these errors are systematically recorded and represented on a graph, they create a performance graph. The collection and organisation of data, and the creation and interpretation of the graph, are the data handling areas that can be developed alongside the measurement ideas of speed.

Estimating lengths

This idea is similar to *The 40-second walk*. It is an indoor activity, and is concerned with both estimation of line length in millimetres and with the challenge of creating a meaningful report when there are several pieces of data about several different individuals. The graphing and interpretation become a major challenge in this situation.

Hit the spot

This simple angle-estimation game for two players uses the computer to measure and store the errors of both players for the five games offered. These results are then presented back to the players, including a graphical form for analysis so they can debate evidence of improvements.

The phone book game

This is a very simple, enjoyable, yet concept-oriented activity. It has features such as estimation, problem solving, pupil generated problems, small group work and friendly competition. Using a phone book, or other thick book with hundreds of pages, students in pairs can play a simple fraction and number game. They challenge each other to open the book at a particular page number. It is easy to keep statistics on how close each guess is, and then make an inference from the statistics as to whether the players are improving.

Other ideas for similar development

1 Estimate angles
2 Estimate lengths outdoors
3 Estimate 60 seconds
4 Estimate length of string in jar (or number of jelly beans)

Break your own record

Children engage in regular physical activities that allow their improvements (or otherwise) to be recorded by counts or measurements. By graphing they depict the changes from day to day, and they interpret the graphs, telling stories about the past and making predictions about the future. These predictions are then tested. This activity is particularly suitable for junior primary students.

Summary of the stages

1. **Choosing the skill**
2. **Conducting daily trials**
3. **Examining the daily results**
4. **Graphing one aspect of daily results**
5. **Examining the results for trends**
6. **Predicting and checking future performance**
7. **Discussion: what have we learned?**

1 Choosing the skill

Students chose a physical education skill which was challenging to them.

The aim was to practise the skill regularly to improve performance. Many children changed their ideas as they initially chose something too difficult or too easy.

Younger children might choose bouncing a tennis ball, or jumping over a skipping rope. Older primary children might choose juggling three small balls or skipping 'double dutch'.

2 Conducting daily trials

We started daily trials. Ten trials per day are enough. Each trial was counted and recorded by the children.

For example, skipping over a rope eight times was a score of 8 for trial 1. You would then start trial 2. Counting accurately was difficult for younger children. Some found it easier to work in pairs — one counting and recording for the other.

Some activities were difficult to quantify. Ben (aged 8) and Leigh (aged 9) kicked and marked a ball between them. It took a while for them to agree that each mark should count as one until the ball was dropped (no matter how inaccurate the kick).

3 Examining the daily results

Range

Each day, the children collected data. Even the youngest children knew their highest and lowest scores. Older children tried to work out their daily *range*, which generated strategies like subtracting the lowest score from the highest score.

Total score

All children found their *total* score each day. Younger children used a calculator; they became adept at reading three digit numbers. Older children used various quick addition strategies (for instance, combining in groups of ten) to estimate an answer prior to checking on the calculator. This gave a reason to constantly practise addition.

Mode and median

All children could find the *mode* (the most common score). There were problems here which children resolved in their own ways. All children could find the *median*, which was the middle score when all the scores were put into order. (If the fifth and sixth were different, we took a number halfway between the two.)

Average

Older children calculated their daily *average*. There was much discussion about what this meant. They grappled with lots of division using concrete material. They talked about what to do with remainders. They used calculators and examined decimal fractions. They 'rounded off' averages which had a fractional result — is it possible to catch a ball 4.75 times?

Midpoint

Children explored putting one day's range of scores on a number line and then trying to find the *midpoint*. This generated strategies like halving the range and adding it to the lowest score. Younger children used strategies like using left and right hands simultaneously to 'jump' towards the centre point.

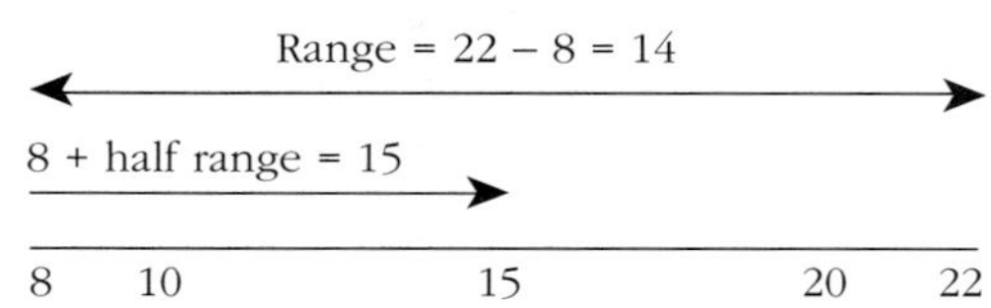

4 Graphing one aspect of daily results

Children chose a feature of their data to plot: daily total, daily average, daily mode, etc. Older children constructed line graphs by hand. Developing an appropriate scale on the *x*-axis was difficult; some children used a computer to generate a line graph which they then read and interpreted.

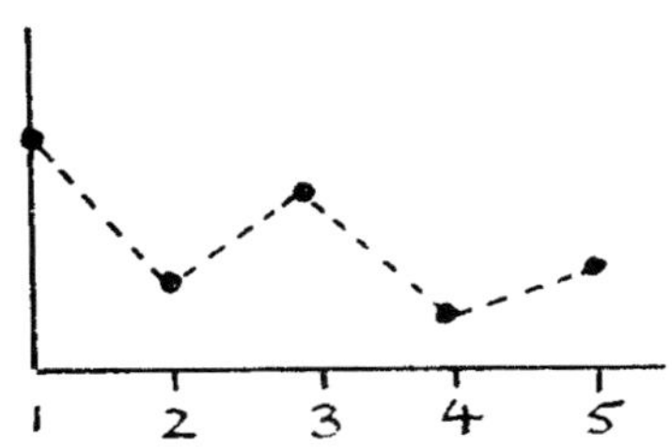

5 Examining the results for trends

This caused some hot debate, because results were affected by so many things. If your average decreased, but your highest score increased, had you improved or slipped back?

6 Predicting and checking future performance

The children saw patterns in their statistics. They knew that if their lowest scores were always between 5 and 10 (say 6, 5, 6, 10, 7) then a logical prediction would be within this range. They made predictions from their numerical scores, and also from the look of their line graphs.

They were asked to make a firm prediction about one aspect of their results before the sixth day of practice.

There was a general strategy of selecting a number within the established range. Some children selected above the range if they saw a pattern of improvement.

For many children, there was a degree of 'accuracy' to their predictions.

I stressed that the predictions were to be genuine, with all the risks implied by that. They shouldn't try to engineer their results to match their predictions. We tried to forget our predictions while we were outside collecting data.

How accurate do you think your prediction will be?

The children appreciated that their predictions were not certainties, but were closer than guesswork.

What type of score is easiest to predict?

Kelly (aged 11) thought it was easier to predict her lowest score for catching a ball than her highest, because it didn't change around as much (less variation).

What affects your results?

Time allowed, fatigue, improvement in skills and concentration were all mentioned by children as important variables.

7 Discussion: what have we learned?

I felt that it was important for children to put into their own words what they had learned. I also had them write a story of the investigation for their parents, placing emphasis on the mathematical things they had been able to use.

How have we used mathematics to help us understand our performance?

What mathematical and other skills have we been using?

What mathematical and other ideas have we been using?

The 40-second walk

This activity concentrates both on ideas of speed and on estimation. Graphing shows students whether or not they have improved.

You will need watches and markers; students need graph grids and Worksheet 1.

Summary of the stages

1. **Posing the challenge**
2. **Gathering individual data in groups**
3. **Discussing results**
4. **Graphing data individually**
5. **Interpreting the graphs**
6. **Discussion and report**
7. **Consolidation**
8. **Extension**

1 Posing the challenge

It is worth taking students outdoors to pose the problem. I chose a tree about 50 m distant and challenged students to walk there and back in *exactly* 40 seconds.

There is a trap. As soon as you take your first step, you are committed to the speed you have chosen; no speeding up or slowing down! When you get back to the start, the timer will tell you how long it took.

We'll have five attempts in all, and after each I would like you to record the time it took and your error.

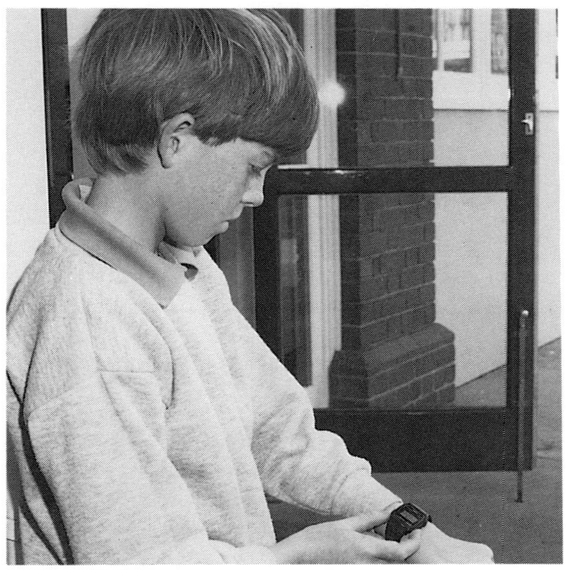

After one demonstration, I asked students to form groups and decide how to conduct their measurements and record data, sharing the tasks around the group.

2 Gathering individual data in groups

Students worked in groups of six. Each group marked a walk of about the same distance on a different part of the oval. Students took turns to be the timer, and all students walked *individually*. They each did six walks.

The activity was a highly successful means of collecting data. The students were intensely involved; it was obvious they felt ownership of the task.

The very low ability children could not grasp the idea that it was not a race. They felt they had to come back first, even though they were too fast.

I found it better for each graph point to be plotted straight after each trial; then the graph and its meaning emerged over several trials. Some schools left graphing to the next class session.

3 Discussing results

The informal discussion of speed, distance and time relationships that took place here was at an intuitive level.

The real value seemed to be in the informal discussions within the group as they physically experienced the relationships.

Students readily recognised that if their time was too low, they would need to slow down.

I always use this activity prior to introducing formula to Year 8. Their personal visual imagery seems to help a lot in making sense of the algebra.

4 Graphing data individually

There are two dimensions to this activity. One is the conceptual understanding of measurement concepts of speed, distance and time. The other is the use of statistics to monitor their own performance.

How could you represent your results visually to show if you improved at estimating how fast to walk?

Time

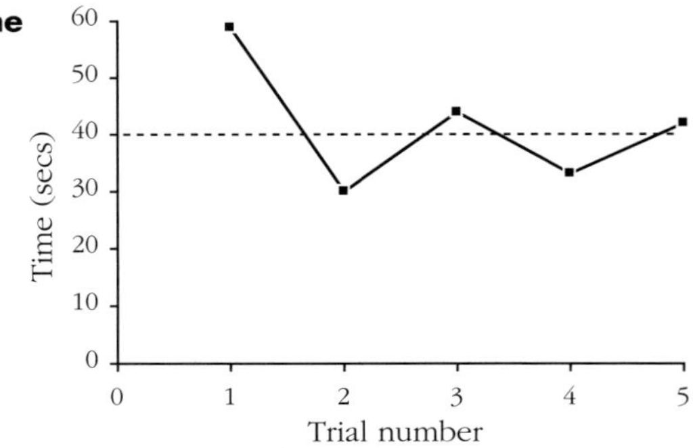

Most students quickly thought of graphs, both of time and error. They were keen to make both graphs to judge which would show the most information. They discussed the relative merits of bar and line graphs.

Error

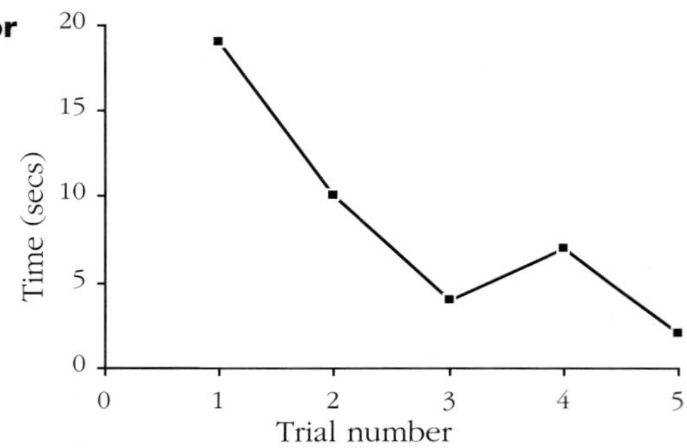

I gave them Worksheet 1, which they used for plotting both errors and time.

The experience of collecting data was vivid in their minds so the meaning of the graphs was evident.

5 Interpreting the graphs

After some group rehearsing of how to explain the graphs, I had several students sketch their graphs on the board and then tell us what they thought the graphs showed.

Share your understanding of what your graph shows with others in your group.

After discussing it in groups, I asked them to write down their preference for the error graph or the time graph and give their reasons.

There was quite a lot of discussion about this, which showed me just how hard graphical interpretation really is.

I realised I could learn much about how they think by listening to their questions and answers.

6 Discussion and report

I asked them to tell me all the mathematical things that they had been doing, and I made a list on the board.

They wrote a brief report of the activity for their parents, explaining what they had done and the mathematical skills they had used.

Their suggestions ranged from reading the time, to measuring, recording information in tables, drawing graphs and interpreting the graphs. They had also worked on problem solving in groups, made and interpreted a mathematical model (the graph is a model of the learning), and held mathematical arguments in which they tried to justify their interpretations to others.

7 Consolidation

On the bottom of Worksheet 1 are problems about walkers with their times and graphs; I asked students to work in pairs on it.

These questions are open-ended and interpretive, yet still have all the mathematical issues of accuracy and understanding.

Although Worksheet 1 questions looked traditional, students had a strong visual grasp of the situation and were able to work out the answers from first principles because they had physically experienced the walking.

Some younger students could not manage question 3. The notion that a shorter time means going faster was beyond them.

Students were quick to point out a difficulty they had not noticed with their own graphs. The size of the error does not tell you whether it is over or under; they had the extra information of the table to help with their own results. In question 2, they could not tell whether the walker was increasing or decreasing speed after each result.

8 Extension

Influence of groups

In the MCTP activity on which this is based, it is suggested that students walk in groups of about six at the same time. I thought that it was possible for students to be influenced by the group, so we walked alone. However, when I put the idea of testing the extent of group influence to the students they were keen to try it and see.

Each group did six walks while I timed them and provided feedback on their individual times. Then they graphed their own times and compared them to others in the group. Then they compared the graphs of their performance individually with the group performance.

They were amazed to see just how much influence the group had on their walking speeds.

Estimating lengths

Students need graph grids and Worksheet 2.

Summary of the stages

1 **Posing a hypothesis**

2 **Testing the hypothesis**

3 **Graphing and interpreting data from the group**

4 **Class results**

5 **Graphing and interpreting data from others**

1 Posing a hypothesis

I spent time with them on the idea of proposing a hypothesis to test.

Do our skills improve with feedback and practice? Do you get better at estimating lengths of lines?

If skills really do get better, then what can we expect to happen? How could we test this?

A hypothesis is just a specific statement of what we think will happen as a result of something. It must be able to be tested to be useful. This means that we can carry out a test to find if the prediction of the hypothesis is true or not.

2 Testing the hypothesis

The class used Worksheet 2 to test themselves.

After estimating each line length in trial 1, they were told its measurement and calculated and recorded their error. This provided feedback to inform their next guess.

The students were excited by their improvement in performance. They found the challenge to improve their performance compelling, and gained great satisfaction from success.

3 Graphing and interpreting data from the group

The groups were given the task of deciding how best to represent the results to discover whether or not there was any improvement in their ability to estimate. This was treated as a *group problem-solving task*. They drew graphs, found averages, etc.

When they had finished, they compared the different ways of treating the data by the groups. A discussion led to an evaluation of the methods.

The students worked in pairs. They analysed and compared their own performance with that of their partners.

Some schools reported that students were sensitive to the fact that errors tended to be smaller for short lines than for long ones. This led to a consideration of percentage error:

$$\frac{\text{error}}{\text{length}} \times 100$$

4 Class results

I suggested that while individual performances vary greatly, a more significant pattern might show up in class results.

To test this idea, we collected all the individual errors for each line on the board. Students simply came out and wrote six numbers in the right places. Then we all worked out the averages, and compared results.

It took time to collect the data but it was well worth it. The data belonged to the students and everyone was involved.

> Students got plenty of calculator practice working out the averages.

Finally, after the inevitable checking for accuracy, we drew the graph of these six averages.

> It was very illuminating — everyone felt good about what the graph showed.

Average of class errors

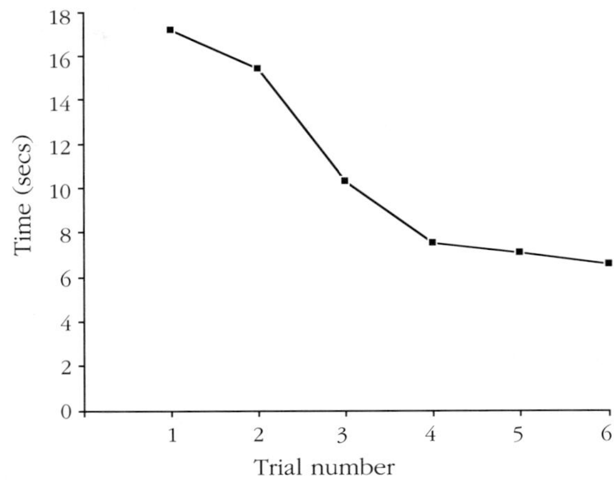

5 Graphing and interpreting data from others

I asked the teachers in my school to try estimating the lengths of the lines in trial 1 on Worksheet 2, using the same process. As a result, each teacher provided six pieces of data for the students to analyse.

For interest, the data is included below. Your class might like to compare their performances to those of these teachers. They might prefer to do it with the teachers in your school.

Line	Joan	Sue	Nev	Phil	Jayne	Ho'd	John	Ken	Chas
1	50	50	80	80	85	70	60	60	66
2	60	75	65	65	70	70	72	61	61
3	100	93	95	94	90	110	85	93	83
4	52	54	58	38	55	55	45	51	58
5	50	48	52	60	55	50	40	55	49
6	85	76	80	80	63	92	75	83	81

The students worked hard to produce graphs and descriptive statements which were presented to the staff. The demand to produce a report to a deadline and to actually communicate it to someone else was met by most students. Several came into the staff room at lunchtime to explain in words what their graph was about, and why they decided to graph it that way. This was a real challenge, and was done very well.

> The *hardest* part of solving a problem is *working out exactly what it is that you are trying to do*. This is particularly the case in the area of statistics, where the purpose of the original activity can become lost in a wealth of easily produced data.

> I believe that to tell the students exactly how to do this work, and not leave most of these decisions up to them, would be to spoil a chance for them. They need to learn to grapple with the challenge of how to best represent data so that the graph tells the story you want it to tell.

> There were very different levels of skills both in grasping the essential problem, namely, what it was that had to be communicated, and in the graphing and graph interpretation.

Hit the spot

This simple activity involves two players estimating angles on a computer. The major aspect is that a computer is used to measure and record errors. These are subsequently presented to the players in graphical form for analysis. The program is on the *Chance and data* disk, if available.

1 Two-player game

The activity has two purposes:

- to practise angle estimates
- to use statistics to monitor improvements.

I demonstrated to students how to play the game. By estimating the angle upwards from ground level each player aims at a spot that moves to random positions around the screen. The difference in angles between the 'correct answer' and the student's guess is the error, and is recorded in a table and used for the error graph.

The game aspect of two players added great motivation.

2 Print-out and analyse error data

After five estimates are finished, students obtain the errors and graphical report. They use this to debate the level of improvement, and to write a summary of what these reports show.

The debate is often very lively as both players seek to convince the other that they have improved the most.

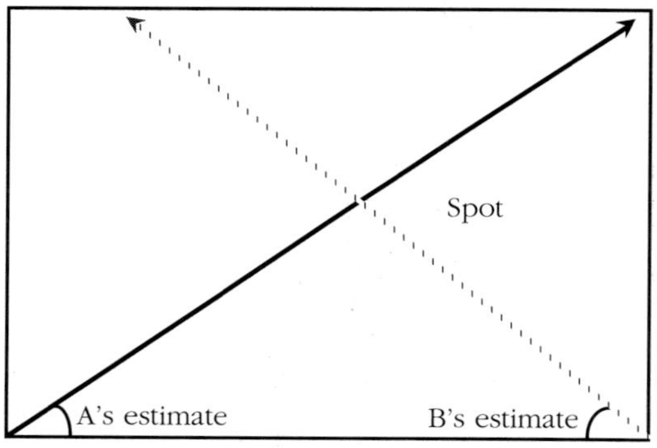

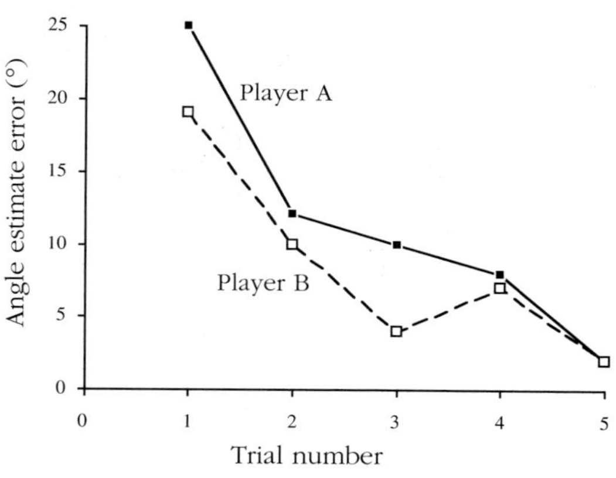

The phone book game

A phone book or other large book with hundreds of pages.

Summary of the stages

1 **Posing the challenge**

2 **The game: gathering and graphing data in groups**

3 **Discussing and interpreting the results**

4 **Discussion and report**

5 **Extension**

6 **Consolidation**

1 Posing the challenge

I demonstrated the game to the whole group.

This phone book has 1247 pages. Open it as close as you can to page 388.

Any large book with hundreds of pages can be a substitute for the phone book.

2 The game: gathering and graphing data in groups

Ask pupils to play in pairs. One player creates the challenge for the other. Each player records their guess and the error on a simple worksheet.

Page no.	Guess	Error
411	388	23

After about five games, ask pupils to graph their errors for each game.

I asked the students what maths they thought was involved; some said fractions, one said ratio, some said numbers and another said length.

3 Discussing and interpreting the results

What *picture* is painted by the results, either in the table or from the graphs?

Teachers favoured an open question initially such as 'What do the figures tell you?' rather than the more directed 'What is your average error?'

The main question is clearly 'Are you improving?' and 'How do the results support this?', but all sorts of other questions pop up. One pupil said 'You make more mistakes when the number given is larger!' So we asked everyone if they thought this was true, and then the class designed a way of using the results to find evidence. (We agreed at the end that it was true!)

Phone book errors

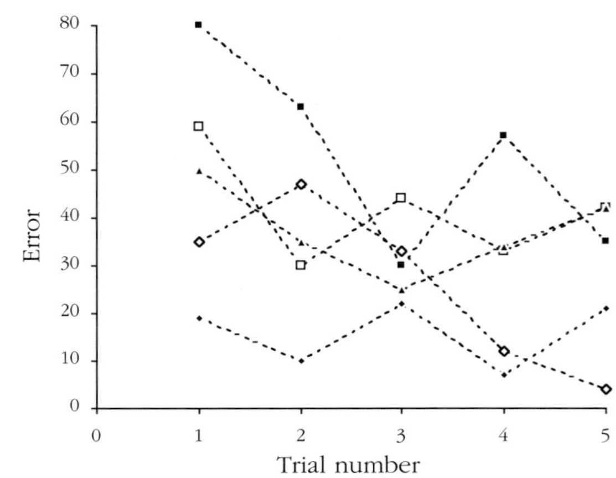

4 Discussion and report

The final challenge was to write a short report. It was to be written as if it was their maths report for the school records. They had to justify any claim they made using statistics or graphs.

5 Extension

Using your results can you predict how 'good' you will be after another five trials?

This question asks pupils to recognise trends and to make predictions from these. The predictions can then be easily tested.

There was heated debate about whether the results of trials 9 and 10 fitted our prediction.

Phone book errors

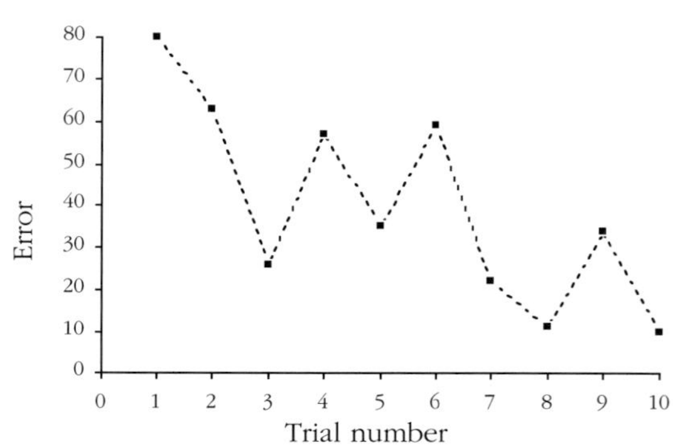

6 Consolidation

Does the improvement hold up over time? Repeat the experiment two or three times over successive periods. Then compare the results. Could you claim that the improvement is maintained?

We took a class average of all the errors for each day and compared these. The class were delighted when it showed a very real class improvement over the three days.

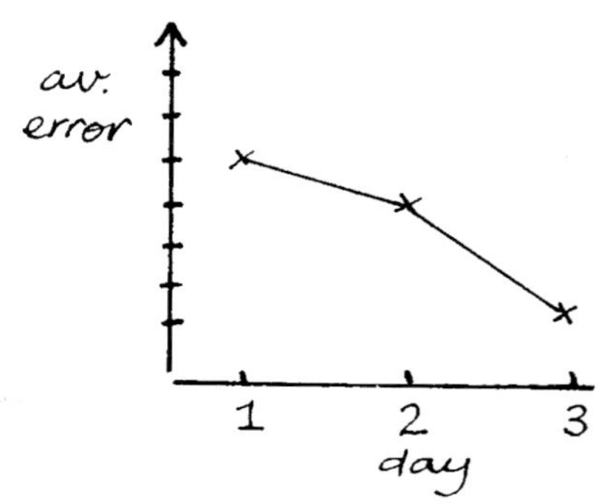

Content epilogue: background information for teachers

These five activities act as models for similar situations where data collection can lead to graph drawing and interpretation. In *Break your own record* the emphasis is just on the *performance* data. It may be a number (of skips), or a distance (of a jump), or a time (for a walk), etc. There are no errors because there is no 'right' answer. In the other activities the emphasis is on *estimating*: a speed (in order to get a time correct), a line length, an angle, or a ratio. The data or errors will be recorded in a simple table, and can be graphed by students either directly on grid paper, or by the simple use of a computer graphing program to draw either lines or columns.

There is a specific set of steps that has been used as part of these investigations. Here is one version of the process.

1 Try to understand the problem.
2 Pose a hypothesis to test (for example that trial 6 will have less error than trial 1).
3 Design an activity to gather the data, and carry it out.
4 Represent the data graphically or in summary form.
5 Interpret the graph (or summary) saying clearly what it tells you. Write a report.
6 Suggest new things to explore as a result of what you now know.

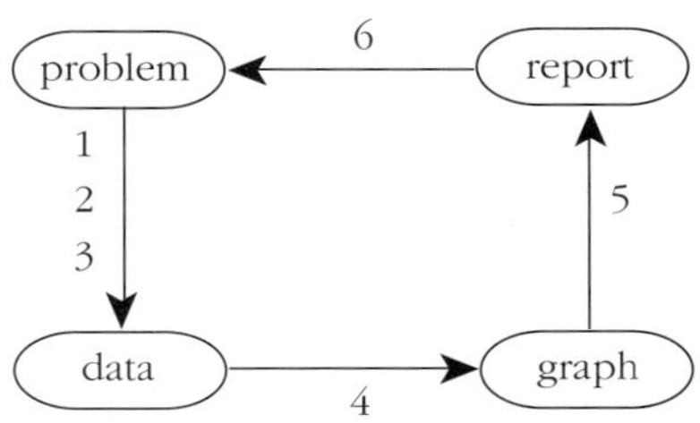

Trends

From five scores it may not be possible to make much of a prediction. It all depends on whether a trend is fairly obvious. A trend is simply a general kind of pattern. But real values often go up or down in erratic ways so that a 'pattern' is not really obvious.

Straight-line trends

The easiest trends to see and to use for predicting the future are straight lines. The numbers either rise or fall by the same amount each time. Very few cases will match this simple picture, but you may want to say that the data show a steady improvement, even though it is not quite as clear cut as a constant change each time. If you can sketch a line through the points, it is often called a 'good-fit' line. If you continue this line on into the 'future', you can use points on it to predict future numbers.

The other 'steady' graph is one which seems to stay about the same all the time; it has no obvious change. The best prediction from this is just the average of the previous values.

Graphical interpretation

The skills of choosing scales and plotting points are essential parts of the drawing process, but in the interpretation stage it is more important to concentrate on the overall shape of the graph: the story it tells as a whole. Students benefit greatly from a chance to discuss the meaning of a graph; many different ideas are likely to come from such an open discussion, and these will reveal much to you about the way your students are thinking about graphs.

Having a clear understanding of the problem and what it is that is being tested, and writing a report to a particular audience, make drawing a graph, interpreting and writing about it a more purposeful activity.

Other statistical measures for measuring or comparing

In addition to line (or column) graphs to show stages in developing a skill, we can use other statistical measures to do this. For example, the *errors can be averaged* to compare performances of two learners.

Jan's errors	20	15	5	3	7	Average error = 10
Bill's errors	15	5	10	6	4	Average error = 8

Percentage errors take account of the fact that less absolute error can be expected from a shorter measurement than a longer one. Calculators are the best way to work these out. For example:

If the error is 5 mm in a line length of 25 mm
then the percentage error = 5 ÷ 25 = 20%

Worksheet 1: The 40-second walk

Trial	1	2	3	4	5
Time					

Trial	1	2	3	4	5
Time					

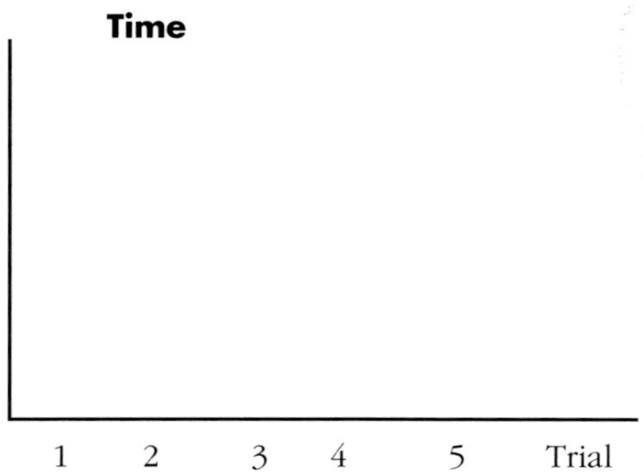

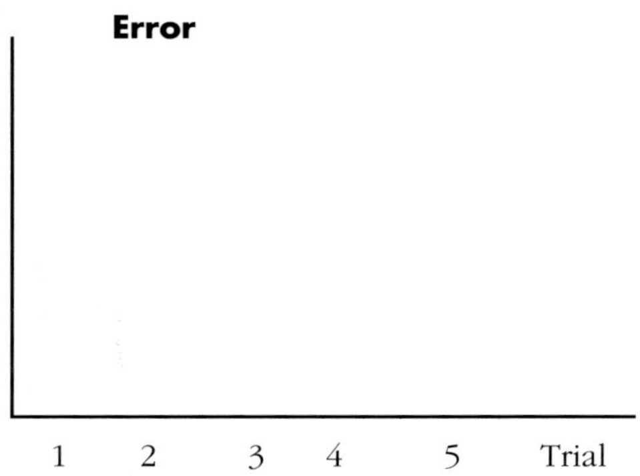

Some problems

1 Here are the data of two walkers. Who do you think has improved the most?
Walker A: 32, 42, 36, 45, 41 Walker B: 30, 35, 38, 41, 45

2 What story does each of these graphs tell?

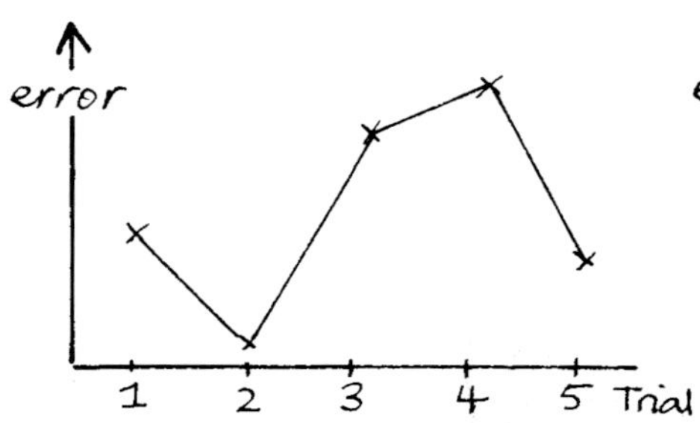

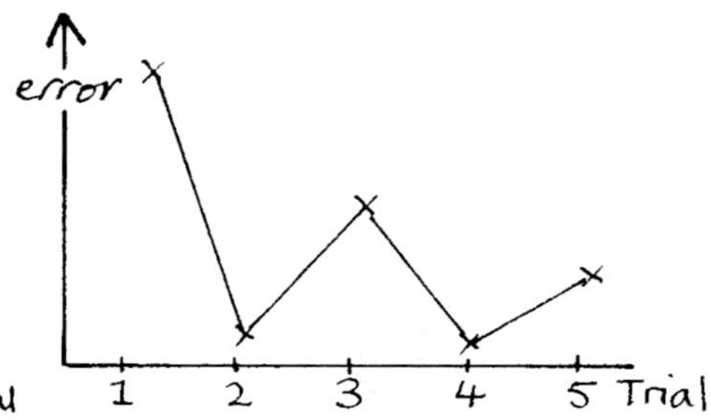

3 A walker took these times to walk from the classroom door to a tree.

Trial	1	2	3	4	5
Time (secs)	56	44	39	42	40

a In which trial did the person walk the fastest? _______ The slowest? _______

b Put the trial numbers in order of walking speed, slowest first.

c Sketch a graph of these speeds in the order of trial numbers.

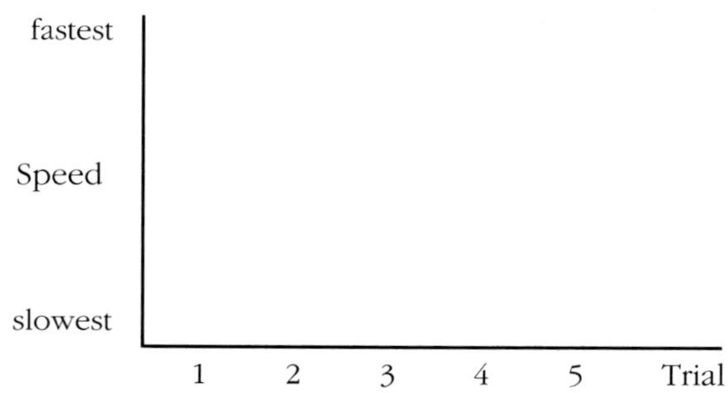

Worksheet 2: Estimating lengths

Trial 1

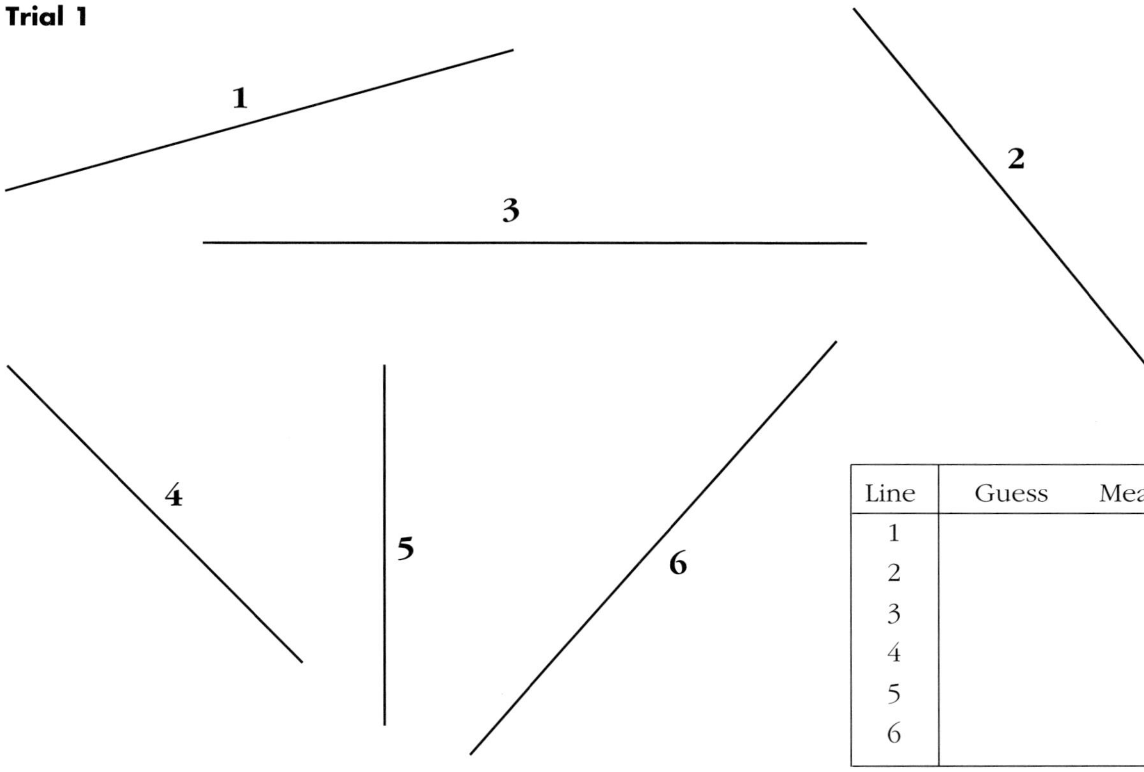

Line	Guess	Measure
1		
2		
3		
4		
5		
6		

Trial 2

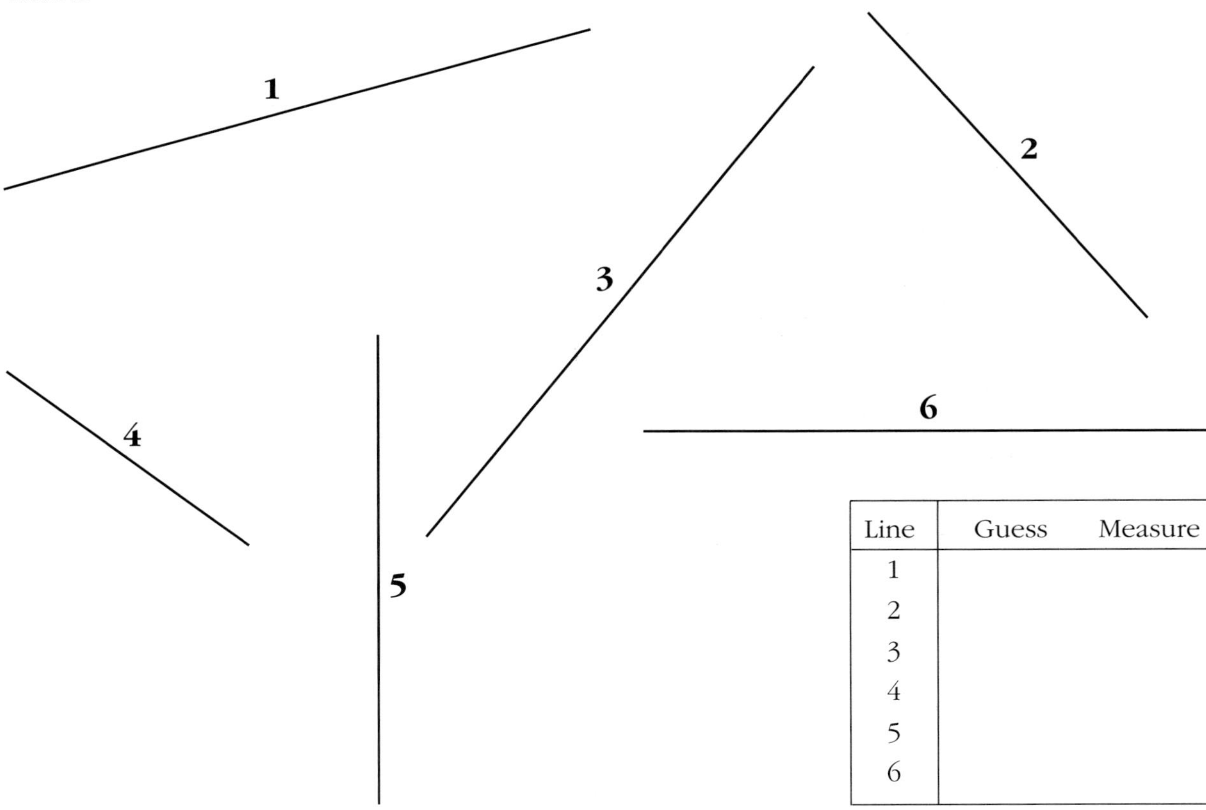

Line	Guess	Measure
1		
2		
3		
4		
5		
6		

Soup kitchen

Years 1 to 6

Children use data collection and graphing to arrive at a recipe for soup which includes only the ingredients liked by the majority of children in the class. Children compare the results of an open-choice questionnaire with one where preferences are from a restricted list of ingredients. The soup can then be made.

Features

Teaching and learning

- Using real data for a practical purpose
- Comparing the results of an open-choice questionnaire with one of limited choices
- Linking mathematics to science, social science and language
- An active conclusion based on the mathematics used in the investigation

Major mathematical content

- Developing an understanding of the mathematical ideas associated with terms such as *majority, minority, most* and *least*
- Collecting and interpreting data
- Choosing criteria for processing data
- There are many opportunities to develop content from other strands, particularly measurement and number

Mathematical content is further discussed in the *Content epilogue*.

Equipment and preparation

- Be prepared to actually make the soup from the recipe you create in this investigation, even if you do not invite parents in or develop a major project

Expecting timing

- Four or five sessions before the actual buying, preparing and eating

A teacher's story

This sequence was the culmination of a science and social studies integrated term program for my Year 3 class in which we studied the growth and characteristics of vegetables, and the production chains for various foodstuffs in supermarkets and restaurants.

For the final project we decided to set up a lunch-time restaurant for parents with a very restricted menu — namely soup and toast. We began by talking about various types of soup and the ingredients we might find in these. I asked the children to individually list all the ingredients they liked in soup. The result of this open survey was considered and the list of ingredients was modified.

Children then voted for each item so they could make up a list of ingredients liked by the majority of children in the class. I tried to find ways of helping them understand exactly what 'majority' means. We also gave family and other staff the opportunity to vote on our list of ingredients and compared the results to our class results. Our final recipe came directly from the class results.

I felt the children gained an understanding of the mathematical notion of 'majority' as well as an idea that polls could be conducted in different ways. They realised that how the list was devised and how the votes were processed could affect the final choice of ingredients in our soup.

This description is of activities which specifically develop the mathematical ideas of chance and data. Learning is also enriched by the practical side. We actually purchase the ingredients at the local super-market and the soup was made according to our survey results. I feel that more effective learning opportunities were provided by establishing this real purpose. (It also helps to motivate the children.)

Comments

This investigation proved to be enjoyable, interesting, and also relevant since the weather was cold.

High interest was generated by the goal of producing an edible conclusion.

I soon realised the value of using real practical mathematics in class. The structure of this investigation is also adaptable to other topics.

Summary of the stages in this investigation

1 **Initial gathering of data**

2 **Making a limited-choice questionnaire**

3 **Deciding on the most popular ingredients**

4 **The concept of 'majority'**

5 **Applying 'majority' to class size**

6 **Choosing ingredients**

7 **Extensions**

1 Initial gathering of data

Let's find the best things to put into a soup. What do you like most in soup?

My students were simply given a sheet with the question: Write down all the things you like to eat in soup.

Through the social studies topic, and because we had cooked in class previously, the children understood the term 'ingredients'.

The rather strange mixture of ingredients that appeared made me think that maybe we should decide first what kind of soup it is, and then what goes into it.

I forgot about chicken; I only wrote vegetables on my list.

I didn't think of noodles, they're really yummy.

I want to add noodles to my list.

Out of class time I listed all the soup ingredients mentioned by the children on a large sheet of paper. In a whole class session we went through each child's written list and used the usual tally notation to record the frequency of children choosing each ingredient.

The aim at this stage is to expose the children to the idea that very disparate results are often obtained in this type of survey. Its use is in providing a wide range of information. However, the findings might not be accurate because they depend on memory as well as preference.

This initial 'open-choice' tally showed several ingredients which were only mentioned once, such as spinach and bean sprouts. There were also a number of popular ingredients, such as chicken and carrots, which were mentioned by most in the class.

2 Making a limited-choice questionnaire

I wrote the shorter list of ingredients (wanted by at least two people) on another second large sheet of paper and asked the class to vote.

We checked each ingredient and if no extra person thought it should be included, then the 'one vote ingredients' were eliminated.

It's a long list. How can we shorten it?

Let's cut out all the ingredients only one person wants.

How many ingredients can we vote for?

I like all of them.

If you aren't allowed to vote for all you like, we won't really find out the most popular ingredients.

I wanted children to understand that there are a number of ways to go about setting up a poll. A poll has to be planned to suit what it is we want to find out.

Some chose to graph from the most popular to the least popular, others grouped vegetables together, then meats and noodles. These groupings were very helpful in understanding the graph.

Previous maths lessons had developed column graphing skills so that this was a straightforward task for the children. If the class had not had extensive experience with constructing graphs of this type I would have constructed a large class graph as a way of developing this skill.

Should it be a secret vote?

Who's afraid to tell what they like in soup?

Draw a column graph to show a clearer picture of the results.

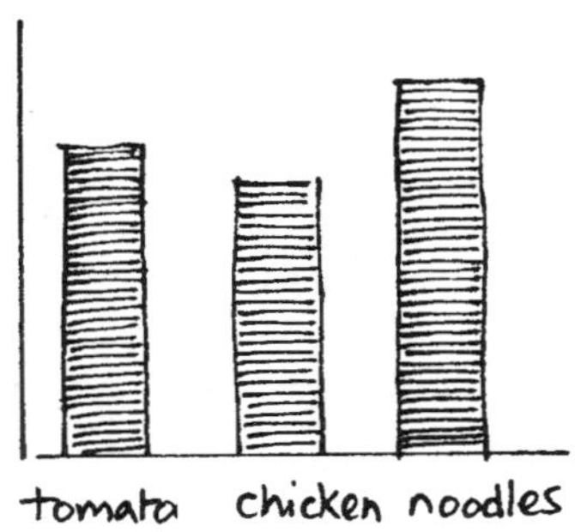

3 Deciding on the most popular ingredients

I suggested we could find out which ingredients the 'majority' of children in the class had voted for, and then see what sort of soup those ingredients might make.

Which are the most popular ingredients?

What about the 'middle' ingredients? Should they be included? How can we decide?

Chicken is the most popular, then tomato.

Very few want cabbage or cauliflower.

Some teachers thought 'majority' was too big a word, and used 'most' instead.

4 The concept of 'majority'

What do you think 'majority' means?

This was a difficult question when presented in this way.

Lots of people like it?

How many is lots?

Most people? More people?

More people than what? How do you decide 'most'?

To develop this idea I asked a group of ten students to sit in front of the class.

How many of the ten would have to stand up for me to be able to say 'The majority of these children are standing up'?

I got them to stand up one at a time and asked the same question after each time.

Are the majority standing up?

The children recognised that five was not the majority, but six was. Seven, eight and nine could also be described as 'more', 'most' or the 'majority' but the terms didn't make sense when all ten were standing.

I used the terms 'minority' and 'less' in the context of the number sitting down. The notion of the majority being in a relationship to the minority was highlighted. We also discussed what you could say if exactly half were standing up. There was no majority or minority, so there was no possible decision you could make.

I chose a different group size (seven) and we repeated the activity.

Activities developed and refined the children's understanding of the meaning of 'majority' and gave it a mathematical context. The idea was used to established a numerical basis for choosing the ingredients favoured by the majority of children in the class.

I challenged one girl to select a group of 16 children in which the majority were girls.

Whereas children have an intuitive sense of these types of situations, such as during their own playground decisions about which games should be played, their awareness of a consistent pattern in relationships has often not been developed. It is important to associate a clear meaning with language used.

Can you prove that if 14 girls were chosen, they are a majority?

B B G G G G G G
G G G G G G G G

We talked about a 'large' or 'small' majority and saw how these terms could make sense only in relation to what's left.

What is the smallest number of girls you could chose?

What if a small majority of the group were boys?

B B B B B B B B B
G G G G G G G

Other children wanted to have a go, so I continued to give descriptions of groups for individual children to display for the class. The child had to arrange a group size of their choice to fit the criteria given. The class discussed the range of possibilities each time.

Why is it a good idea to have an odd number in small groups which have to make decisions?

An important aspect of the use of groups of children to demonstrate this idea was to provide the opportunity for movement as well as seeing concretely the relationships we were discussing. The requirement for physical action kept them focused for a longer period of time.

Once I was sure they understood I gave each a piece of paper and challenged them to work out a rule that would give the smallest number that would be a majority for any number of people.

This was quite a challenge, well beyond what would be expected at Year 3. But they had a clear understanding of what constitutes a majority and what doesn't for any number, so were able to grapple with the problem and experiment with numbers.

Several said 'divide how many by two and add one', or 'one more than half' for even numbers. We tested this for different even numbers until all were satisfied it worked.

It works for odd numbers too. For nine, half is about four (you couldn't have a half of a person) and then add one on makes five. That is the majority because there's only four left.

The children refined their intuitive sense that majority means more than half, but they generally had not considered the minimum size of a majority for a group. The emphasis given to this aspect established a firm understanding of this concept.

5 Applying 'majority' to class size

What is the lowest number of children to vote for an ingredient if we said the majority of children liked it?

We used our rule to establish this and then I had the class line up in pairs to see visually what the actual formula did.

I joined the group myself and everyone lined up again in twos to establish the idea clearly.

We explored smaller groups of odd and even numbers and used our rule of actually pairing up to form two equal lines, if possible, and 'adding one' to make a majority as necessary.

One child wanted me to be in the voting. I asked if it could make a difference to the number that would be a majority and they decided that 16 would still be the majority in our class of 31 (including me). This established the idea of a casting vote when there was a split decision.

6 Choosing ingredients

In our class we had firmly established that 16 was the 'majority'.

Look back at the graphs. How can you show which were the ingredients liked by the majority?

You could draw a line across on 16 and choose the ingredients that reached that mark or higher.

7 Extensions

A: Parents and teachers voting

To find out if the parents and other teachers would also like similar ingredients we decided that each child could take home a copy of the list of ingredients we used for the class voting and ask their parents and family to vote Yes or No for each ingredient. We also polled the other teachers with the same list and collated all this information back in class again. We calculated how many people were asked to vote and worked out the 'cut off' figure as the majority.

We found the majority chose almost exactly the same ingredients as did our class.

If the number for each ingredient was calculated by adding the class vote to the family and staff vote, would the most liked ingredients differ?

There was some discussion about this aspect, some immediately said there would be more ingredients in it, forgetting that the 'cut-off' figure would also increase.

This new calculation was completed and it was found that the original list remained essentially the same.

This may or may not happen in another situation, but the reasons can be a source of discussion and investigation.

B: Preparing to shop

We settled on the ingredients which were chosen by the majority of children in the class.

Groups of students were involved in measuring and calculating quantities for soup, devising a shopping list and estimating costs using newspaper advertisements, so we'd know how much money to take shopping as well as how much to charge 'customers'.

Further opportunities for mathematical development were afforded by these activities.

With parent helpers, we had an excursion to the local shopping centre, each child having a shopping list.

C: Preparing and serving the soup

Other mathematical activities carried out by the children included:

- planning a sequence of work to prepare and cook the soup in time for the luncheon;
- planning a roster of toast making and buttering;
- collecting the money and calculating profits;
- making and serving the soup to parents and staff in our lunchtime 'restaurant'.

It was very important to justify the mathematics by actually making the soup.

The class actually made a very palatable soup which was enjoyed by all.

Content epilogue: background information for teachers

This project is nominally about making a soup. However, it is actually a vehicle for some quite sophisticated ideas about surveys (such as whether the questions should be open-ended or restricted-choice) and about the concept of majority.

Surveys

The idea behind the development of this investigation was a belief that children need to be exposed to a range of choices during data collection so that they develop an understanding that what is asked, who it is asked of, and how the data are processed are all important possible variation which may affect results.

The accuracy and usefulness of surveys depends on good design. Newspapers often report 'survey results' but rarely question the validity of the way surveys are designed, conducted and interpreted.

Ideas specifically mentioned in this investigation are:

- *Open-response question*: You can say what you want, but the response is often dependent on your memory or how much effort you are prepared to make at the time.
- *Restricted-choice questions*: Here you can only vote for choices that are offered, so they need to include all the likely possibilities. Should you be allowed to vote for all the choices offered or just choose a few, say three?
- *Secret or open voting*: Children know that voting in political elections in Australia is normally 'secret'; that is, you can actually vote however you wish and no-one can tell what you've written. Why do we have this system? Is it open to abuse? What about the alternative? Why might people not want to vote in public? Could peer pressure influence your choice of when to raise your hand?

Voting and 'the majority'

Our voting system requires that the winner must have a majority of votes. Language of small and large majority and the inverse ideas of minority are important here. (In a democracy where the ideas of 'the majority' always prevail, how can the rights of 'the minority' be protected?)

At quite a young age children can learn about the need for more than half for a majority. At later ages this can be related to the 'preferential system of voting' where the preferences of the candidate receiving the least votes are allocated to the others until one has the majority of votes. Many countries use the 'first past the post' system, in which the largest vote wins, without having to be a majority.

Why do we use a preferential system here? Is it any better than the other? Is it open to abuse or anomalies?

Chairs and tables: just right?

Years 2 to 6

Do your children wriggle a lot on their chairs because they are the wrong size? Do they bang their legs on tables which are too low for them? Is there one best size for your year level, or should your class have several sizes? Here we let children find out for themselves, and learn some statistical methods too.

Bikes, monkey bars and skeletons **(page 272) follows the same pattern; it is designed for older children and introduces more advanced graphing techniques.**

Features

Teaching and learning

- Group investigations
- Uses real personal data
- Demonstrates the power of mathematics to model reality
- Uses a story-shell (Goldilocks)

Major mathematical content

- Uses a standard process of statistical investigation
- Uses measurement in standard units
- Graphs of each person's data
- Frequency tables, column graphs
- Interpret graphs in terms of the original problem

Mathematical content is further discussed in the *Content epilogue.*

Expected timing

- From three to eight sessions

Not a small man, but a very large chair!

Equipment and preparation

- Take a good look at the chairs and tables in your room. Do the children fit them comfortably?
- Wool or tapes or rulers
- Possibly an overhead projector
- Additional data are available on the *Chance and data* disk

A teacher's story

It all began when I became conscious of how some of the larger children did not really fit too well on the small size chairs in our room. I decided to make something mathematical of it.

The investigation I used follows the standard four steps:

Clarifying the problem
We worked out that the length of our lower leg should be about the same as the height of our chair seat.

Obtaining the data
This meant that the children had to measure each others' lower legs (behind the knee to floor).

Organising the data
We used a graph to show the lower leg lengths.

Interpreting data to answer the question
When they compared the chair seat height to the graph, it was clear that some children had outgrown their size of chair. These children agreed, showing that the mathematics had some power!

Afterwards we carried out an investigation of suitable table heights, using the same four steps, and, as a result of our findings, the school obtained some new tables.

Comments

Children were interested and involved. They could see a practical application for this mathematics in the world around them.

It was very motivating and good fun. However, some of my Year 8 students wondered if this was 'real maths' and remained unconvinced, although I talked about how it is really used in the world outside school. Maybe we have brain-washed them with the wrong ideas about maths.

The reluctant students particularly enjoy these tasks. They don't feel as threatened, they're more interested and they gain a sense of achievement.

It was interesting watching the children discuss and solve something I thought they couldn't do. I was limited and had low expectations.

The investigation made me realise the extent of the maths all round us, and how to draw from it to extend the children past merely measuring.

Summary of the stages in this investigation

1 **What size chairs should we have?**

2 **Clarifying the problem**

3 **Obtaining data**

4 **Organising the data**

5 **Interpreting data to solve the problem**

6 **Extending the data**

7 **What size tables should we have?**

8 **Obtaining and organising the data**

9 **Interpreting data to solve the problem**

10 **Discussion: what have we learned?**

11 **Consolidation**

12 **Extension**

1 What size chairs should we have?

I started this with a little light drama: I mimed the story of *Goldilocks and the Three Chairs*. I didn't tell them the story first; they had to guess it. I used a kinder chair for the SMALL chair, the table top for the BIG chair, and an adult's chair for JUST RIGHT — for me.

The mime included coming to a little house in the forest, knocking on the door, going in and finding three chairs of different sizes.

I sat on the table top for the BIG chair; I dangled my legs to show that it was TOO BIG.

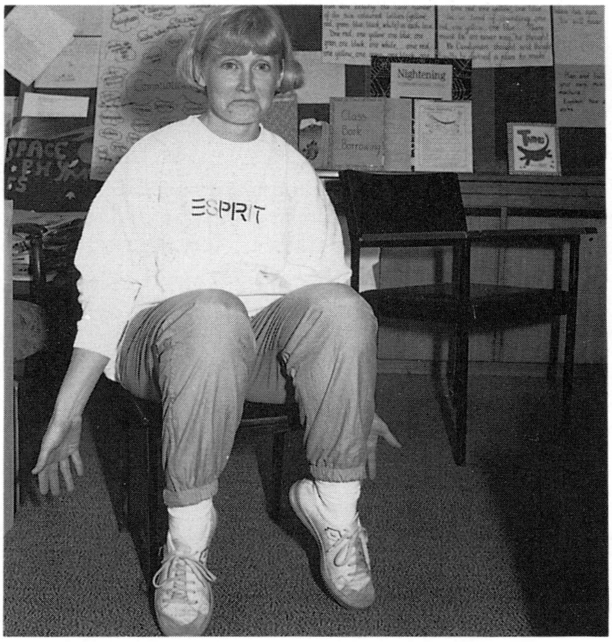

I then sat on a tiny chair. For me this was TOO SMALL.

But for me the adult-sized chair was JUST RIGHT.

I made sure they focused on the sizes of the chairs: TOO BIG, TOO SMALL and JUST RIGHT.

I get a great kick out of hamming it up, and the class enjoyed it! I dropped hints to overcome my deficiencies at miming!

The story was a great introduction; it had all the children's attention. The language was fantastic — simple, clear and precise.

What was the story? Tell me the story as you remember me miming it.

An alternative opening

Take the class for a short visit to the school library. Unless your school is quite exceptional, the chairs there are all designed for the smallest children in the school. Use this fact to raise the issue of the best chair sizes for different children.

2 Clarifying the problem

What is 'just right'? How can we tell that a chair is just right for us? Is the adult chair just right for you? Is the kinder chair just right for me?

Are the classroom chairs the right size for us?

What sizes of chairs should we have?

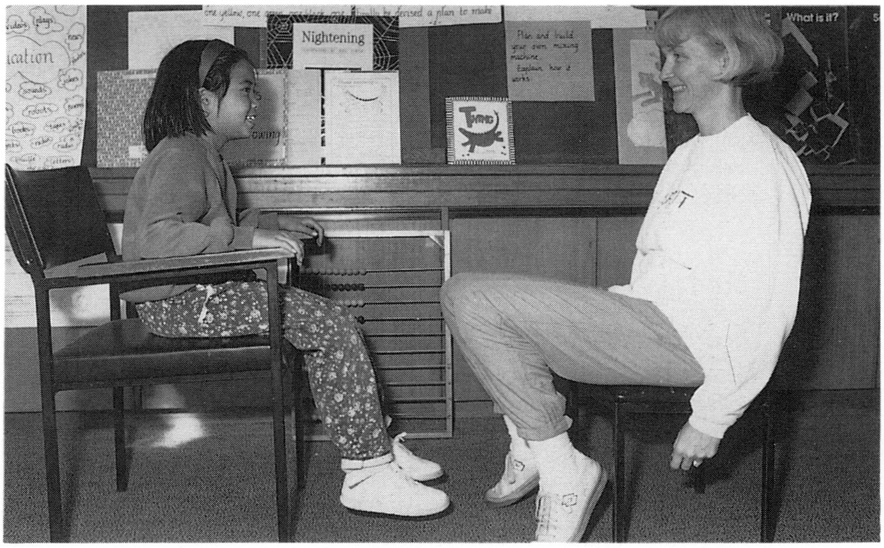

In the discussion that followed I had to give careful instruction on how to sit properly in a chair. (Many didn't really know!) I had tall, short and 'in-between' children sit on a 'demonstration chair'. I also drew stick figures to show sitting straight-backed with support in the small of the back — these helped show the lower leg length compared to the seat height.

> Some teachers reported that although there were chairs of various sizes in the room, they often went to the wrong children!

3 Obtaining data

This led to the need to take measurements of our lower legs (from behind the knees to the floor when seated) and of the chairs in the room. I divided them into groups and gave them appropriate measuring equipment.

> One class used a plank (of about 70 cm) standing upright on which to mark their knee height; when all knee heights were marked on, with initials, the plank was photocopied.

> 'Appropriate' depends on the level of proficiency and understanding of your class. A Year 6 class has no trouble with centimetres, but Year 2 students needed pieces of wool to make exact copies of the lengths being measured.
>
> Be sensitive to the fact that some children may not wish to be touched by others. Let them choose who can take measurements of their bodies.

> When collecting data, children were very aware of whether the measurements recorded were appropriate or not possible. For example: Your lower leg couldn't be...

> **Behind the knee** is better than in front, so that it matches the proper seat height.

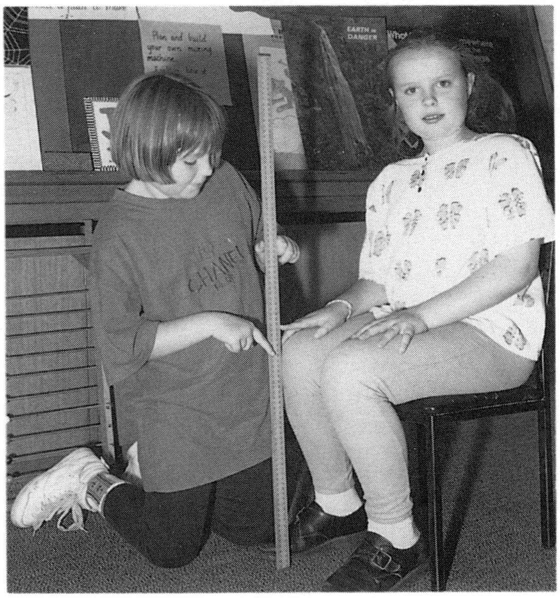

4 Organising the data

At first I had them put their name on the board and next to it the measurements; this made an elementary table, or database. (This would not be possible for children who cannot yet measure in centimetres.)

I graphed the lower leg lengths on the board. Each child's lower leg length was drawn vertically, in order of size, with the shortest on the left.

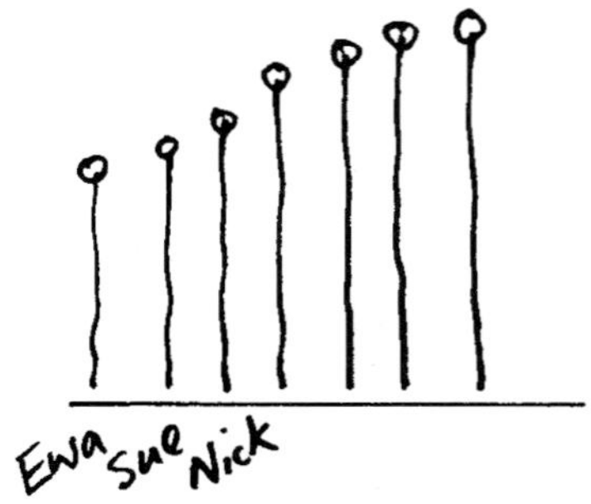

For a junior class this would have to be done on the board. The lengths of wool could be attached to the board in order of size, and names attached.

One teacher had children draw these from the table using a simple scale: 1 mm shows 1 cm. The range of lengths was from 34 cm to 46 cm so they only needed short lines.

One teacher drew these on a graph grid on the overhead projector, one at a time so that all children could say their own measurement and see it drawn. In this way they knew which line was theirs! She then photocopied the acetate sheet for them all.

I used the chalk ledge as a baseline, and children arranged their wool from shortest to longest. Some were inclined to stretch the wool, in order to be bigger than they were!

What seems to be about the 'average leg length' (or the 'leg length that is in the middle')?

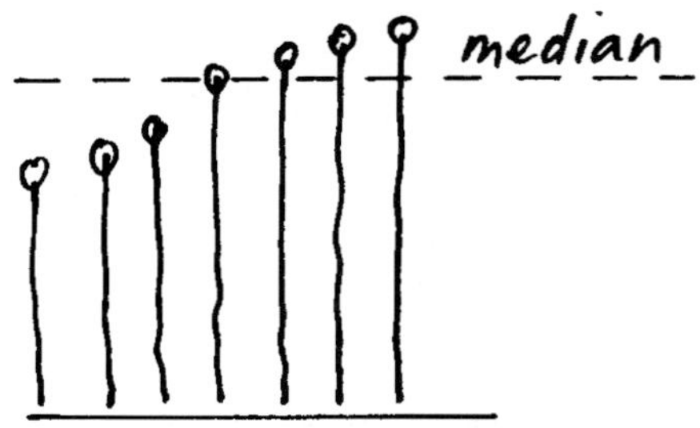

The middle one can be called the *median*; but the name does not matter at this stage. The median may not be the same as the average, but it will probably be quite close, and is easier to work out.

Older children could use calculators to work out the average. Several pairs should do it, one telling the other the numbers to enter; they can then check each other's work.

The next step was to see how the height of the chair seat compared with the leg lengths. We all drew a horizontal line across the graph to show the chair seat height. (In the situation of the graph on the board, it is possible to hold a chair up to the board and compare directly, before drawing a line across the graph.)

Surprisingly, we had seven different size chairs in our room. Children worked out how to represent chair heights on the graph with labelled lines.

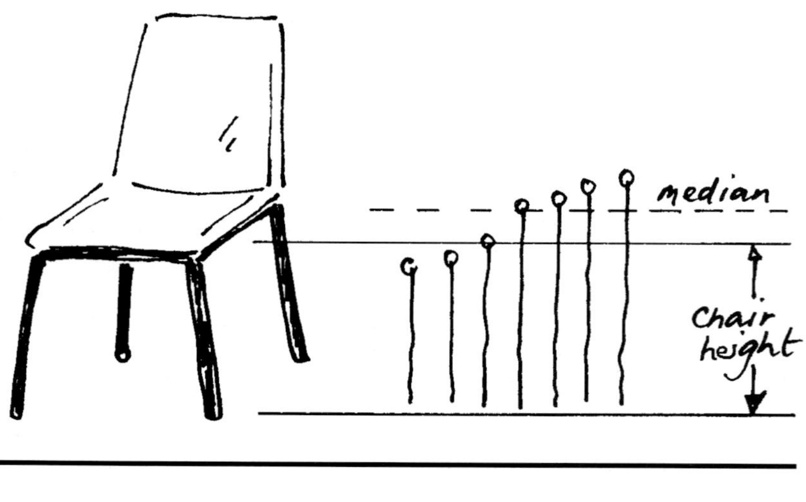

5 Interpreting data to solve the problem

The question arises: how much difference does there need to be between your leg and the seat height for it to be too small or too big for you? This can lead to a complex discussion, but probably is best settled by comparing the prediction with reality.

According to our graph, which children find the chair height 'just right'?

Who thinks the chairs are too small for them?

Who thinks the chairs are too big?

Has the graph been any help to find out?

One teacher felt that the meaning of 'just right' was 'comfortable for sitting' but that this was not the best height for working!

Most children were happy with their present chairs, though the tallest decided that they were unhappy. This is because the chairs in our room are all of the small kind, and unsuitable for the early growers!

One teacher complained that this '...opened up a can of worms. The lesson needs to be treated carefully, or the Principal might have to answer a lot of awkward questions about inappropriate furniture.' Other teachers saw this as an opportunity, and presented a careful report, written by the children, to the administration.

6 Extending the data

At this point the children were keen to find out if there were suitable chairs in other rooms that other children did not want, so they could arrange a swap. We asked several other classes to tell us the lengths of their lower legs (in centimetres), and prepared graphs in the same way. We also asked for the height of chair seats in their room. In each class there were clearly some children for whom the chairs were ill-suited.

The investigation at this stage became a schoolwide matter. Each teacher was asked to decide how many chairs of each size are needed in their rooms. We held a staff meeting to relocate all the chairs in the school!

Children decided what we needed to keep and what we could trade, but we couldn't solve the problem because it disrupted too many classes.

Children were keen to go home and measure the chairs there.

Do you have a special chair at home for your own use?

Are the chairs at home too big or too small for you?

The feedback from parents was positive; they began to realise that maths was useful to them too, and were able to help discuss real needs with the children.

7 What size tables should we have?

The obvious next step was to investigate our tables (you might have desks). To provide motivation, I brought a coffee table to school and placed an adult chair next to it. I also put a kinder chair next to the adult table that I use as my desk.

How can you tell whether the tables (or desks) are the right height? Does it depend on the height of the chair seats?

I invited the shortest child to sit in the small chair and work at the table, another child sat at the adult chair next to the coffee table. Within a few seconds children were asking whether they could measure the tables to see whether they were the right size. (My lot are fast thinkers and take hints well!)

I divided them into friendship groups and gave them 7 minutes to work out a solution to this problem. Reporters from each group later told the class their ideas. An interesting discussion took place.

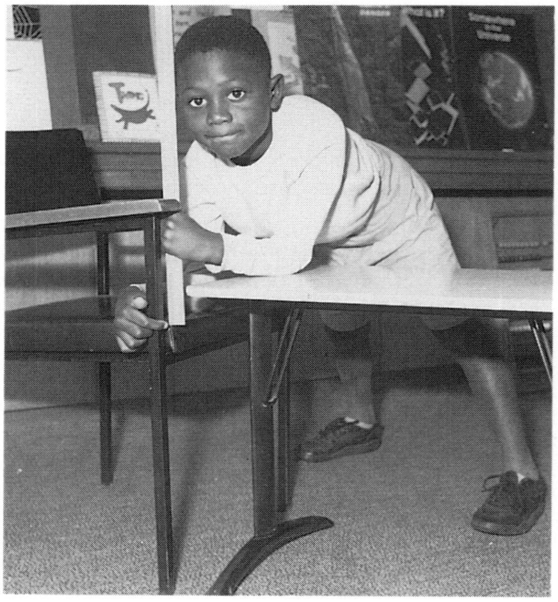

How much higher than the chair seat does a table need to be?

> The idea that table height and seat height should have a common difference may take a while to develop. This needs persistence.

> A number of boys habitually fold one leg underneath them and sit on their lower leg. The reasons for this were discussed.
>
> Many children were unhappy with the comfort of their chairs and desks. Some small desks, with tubes, were difficult to fit legs under. Some children even had sore spots from rubbing on the tables.

8 Obtaining and organising the data

Finding a solution led to groups adjusting the tables to different heights by stacking things under the legs. Different groups came up with different answers, but we found a fairly common agreement and discussion settled the matter.

I let them work on this once they all understood the problem.

> One school used MAB flats (100s) to make centimetre increases in the table or chair heights.

9 Interpreting data to solve the problem

Again the interpretation of the data depends on the amount of tolerance you are prepared to accept. The children made a list showing, for each person, their lower leg length, their chair height and their desk height. They worked out the difference between the last two and decided whether this was acceptable.

Additional data are available on the disk. There are 65 records of children from Year 2 to Year 6, giving their lower leg length measurements.

Are the classroom tables at the right height for our chairs?

Sometimes we push the tables together for group work. What would happen if we had the proper sizes of tables for our room?

Children realised that having different sizes for the desks can be a problem when doing group work frequently.

10 Discussion: what have we learned?

I asked the children to tell me what mathematical things they had been doing as part of this investigation. As they told me I compiled a list on the board. They were amazed at how much they had actually done!

I asked them to write a story about the investigation for their parents, which explained what we did, what we found and what we had learned.

11 Consolidation

I handed the worksheet out to the class. As they looked at the data I asked them to guess the year level of the class; in fact it is a mix of Year 4 and Year 6. Then they completed each part of the worksheet.

This was designed to revisit the major mathematical ideas some time after the investigation. It therefore acted as a review.

I asked them to use the back of the sheet for the first column graph. They chose to use a scale of 1 to 10, that is 1 mm to show each 10 mm.

The two different frequency tables give quite differently shaped graphs. This shows that the choice of the interval for the frequency table is important.

12 Extension

A: Design of chairs

The chairs and tables in your school were chosen from a catalogue by someone. Can you find out who made that decision, and the basis for it?

Some education systems have design standards, or a special limited choice of furniture sizes that allows them to purchase in bulk and save money. Are the sizes of tables and chairs related in these standards?

B: Choosing things that fit

Can you think of other things where you try to buy sizes that fit your body?

How often do you change these items (say shoes, beds,...)?

It is important to let children feel in charge of the method of investigations. In this way they feel that the control belongs to them, and is not imposed on them by their all-knowing teacher. This is part of the process of helping them feel that mathematics can help them make their own decisions. However, you do need to have some ideas ready.

Content epilogue: background information for teachers

Column graphs

The usual use of column graphs is to show the value of some measurement for each thing measured. For example, each student's lower leg length can be shown on a graph which uses a line (or length of wool) to show the length. They may be put into order.

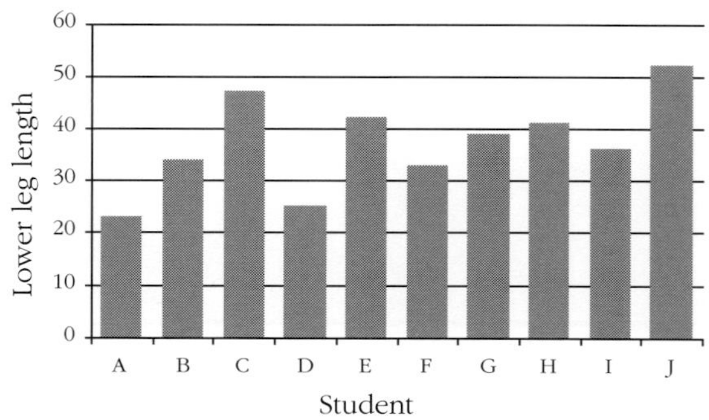

Frequency graphs (histogram)

Instead of showing the measurement, these graphs show how often certain things occur. For example, the number of leg lengths between different numbers of centimetres. In this investigation we try to help students see how the frequency can be seen on the column graph.

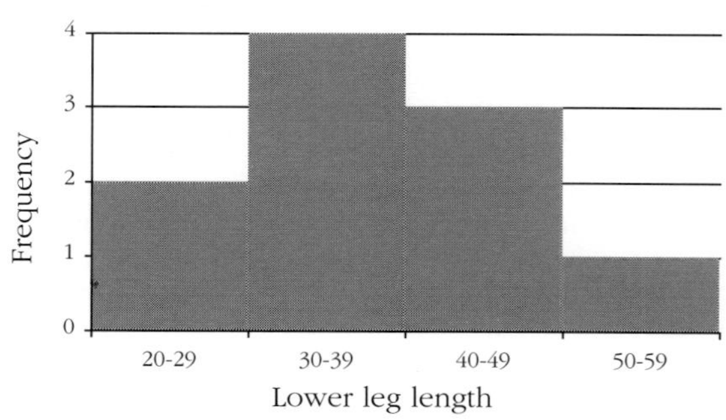

Mathematical models

A graph is a description, using mathematical language and ideas, of some aspect of the world. This is also called a model, as it represents the real thing, in the same way as a toy car represents the real car. If we add a line (or lines) to the graph showing the heights of the chair seat(s) in the room, we have a model that allows us to compare the chairs with the leg lengths directly.

Interpreting the model

What you make of it all is called interpretation of the model. The range of leg lengths we are prepared to say 'fit' the chair height is a value decision that does not use mathematics; it is just a question of how fussy we are prepared to be. Mathematics can only show the mathematical aspects of the situation.

These are the kinds of issues faced by designers of any items that must fit humans, and of course this is particularly important for growing children.

Worksheet

Here are the lower leg lengths in centimetres from a real class.

41	42	43	41	39	38	43	45	40	40
40	46	41	44	41	44	41	43	38	37
43	46	43	35	38	37	45	34	43	40

1 Draw a column graph for this class, with one column for each person. (You could use a computer to do this.)

2 **a** Complete this frequency table.

b Use the frequency table to draw a frequency graph.

Range of lengths (cm)	Frequency
30–39	
40–49	

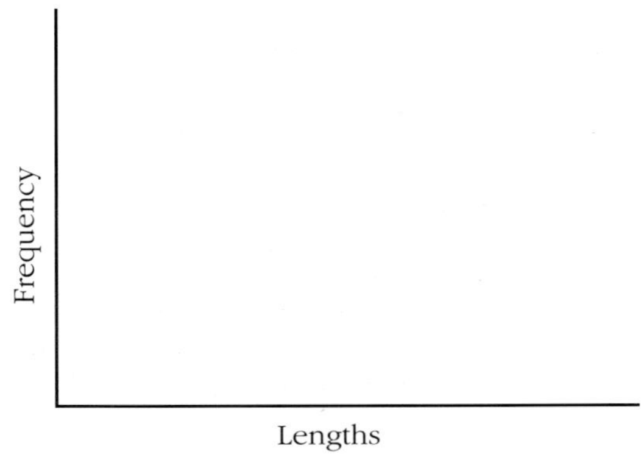

3 **a** Complete a different frequency table.

b Draw another frequency graph.

Range of lengths (cm)	Frequency
30–34	
35–40	
40–44	
45–49	

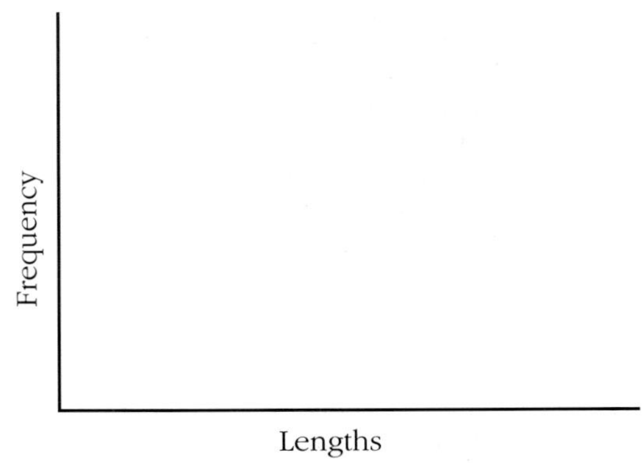

4 Which frequency graph seems to give you the better picture of the way the results are spread? How do you decide this?

5 The chair heights in the room were 37 cm. Comment on this.

Groups and sorting rules

Up to Year 3

Children group themselves and others on the basis of characteristics. Initially these do not overlap, but later some people can have two characteristics at once. These leads to overlapping sets, and children learn ways to represent this situation.

Features

Teaching and learning

- Whole class and small group activity
- Children making decisions
- Working cooperatively
- Active involvement
- Problem posing and solving

Major mathematical content

- Devising rules for sorting
- Identifying and describing both similarities and differences
- Informally representing information
- Learning to model data
- Comparing information
- Intersecting sets, Venn diagrams

Equipment and preparation

- Hoops and attribute blocks (or cards)

Expected timing

- About a week

A teacher's story

These activities focus on the forming of categories and gathering data in order to answer questions which are of immediate relevance or interest to the children. I wanted the children to make as many decisions as possible, within a given framework or context, about what they investigated and how they represented the information. The students were expected to comment or report on their findings (as opposed to reporting on what they did) and use the data to answer questions.

Mathematically the investigation divides itself into two parts, stages 1 to 5 dealing with single sorting rules, and the remainder dealing with two rules about different characteristics. You could consider stopping after stage 5 if your class is not yet ready for two sets that might overlap. However, this works fairly well with quite young children if the problems are put clearly and they have plenty of time to work them out together.

Comments

I found that children enjoyed using themselves as the objects to be sorted. It gave them all a feeling of involvement. When they eventually moved on to sorting other objects, such as the attribute blocks, they showed that they had really understood.

My Year 1 class particularly enjoyed the favourite toy activity. It led to much work in other parts of the curriculum, with measurement, number and language featuring highly.

Summary of the stages in this investigation

1 Using sorting rules to form groups

2 Collecting and recording the data

3 Re-examining the data

4 Groups doing their own sorting

5 Representing information

6 Sorting using rules about two attributes

7 Sorting favourite toys

8 Using sorting rules with commercial materials

1 Using sorting rules to form groups

I started by explaining that I had a sorting rule in my mind that would tell something about the people in the class. Some people would fit the sorting rule and others would not and so we would be sorting people into two groups. The sorting rule would use something that could be seen, like wearing jeans or having blue eyes.

Let me see... Marcia and Julie fit my rule. Would you both stand together over there where we can all see you.

Who else do you think might go in the group with Marcia and Julie?

The children made suggestions about who they thought would fit the rule and as a child was named, they were asked by me to join one of the groups according to whether or not they fitted the rule.

> I like to delay for as long as possible the children giving suggestions for the actual rule.

Once I saw that many of the children knew my sorting rule I asked them to say it, and explain how they knew.

> I feel it is important that children always give reasons for their conjectures.

We played this game a number of times and the children became very confident about how the activity worked.

If you are not already in a group, which group do you think you would belong to? Go and join that group.

2 Collecting and recording data

The children broke into groups of two or three.

Who can think of the trickiest sorting rule?

Each group decided on a sorting rule which they used to challenge the rest of the class.

As each group had their turn, I recorded the number of students who did or did not fit the rule. I tried to use as many different, informal ways to do this as I could.

> Here are some different, informal ways of recording data.
>
> Wearing sneakers: ❖❖❖❖
> Not wearing sneakers: ❖❖❖❖❖❖❖❖❖❖❖❖❖❖❖❖❖
>
> Brown hair: 10
> Not brown hair: 18
>
> Striped shirt: 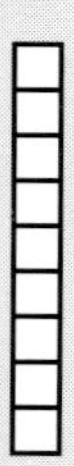 Not striped shirt: 

3 Re-examining the data

Another day I displayed the data that we had collected from playing with sorting rules.

If we made the same groups today would any of this information be different?

Is there anything that you are sure will not have changed? Is there anything you are sure will be different?

The children shared their views, explaining the reasons for their thinking. We actually made some of the groups again and compared the data with that collected during the previous session.

What are some of the reasons for today's numbers being different from what we had yesterday?

4 Groups doing their own sorting

I explained to the children that we would be continuing to gather information that could be seen about our classmates. I also explained that we would be working in groups of four or five to do the work.

I outlined the steps that would be necessary to undertake the task, having the children make suggestions for each step. I listed the steps and the suggestions on the board.

I provided copies of the class list and we discussed various ways this might be helpful in gathering the information.

Decide what to investigate. Do you have any ideas?

- **Eye colour**
- **Kinds of clothing**
- **Number of missing teeth**
- **Wearing glasses**
- **Birth months**
- **Lunch**

Decide how to record the information. How could we do it?

- **Use ticks or strokes**
- **Count**
- **Write down the names**
- **Use blocks**

The children formed themselves into groups. I found it necessary to help some of the groups clarify the characteristics that they were investigating.

Some groups restricted themselves to a characteristic which could be divided into two groups (such as wearing a watch) while others wanted to investigate a characteristic which had several categories (such as eye colour). (Eye colour led to some interesting arguments about what to call different colours.)

> To an outsider, my classroom must have looked like utter chaos when the groups started collecting information. There were 30 children all moving around the room at once! Actually they were very busy and most of them collected the information by observation.

5 Representing information

The next step was to think of ways to show the information that had been collected. The class discussed possibilities and I listed them on the board. The methods I had informally used during stage 2 were quite helpful in suggesting ideas to them.

In your groups explore different ways of showing your information to the rest of the class.

If children are going to be expected to make some decisions then I think it is important to discuss possibilities with the whole class. This is not intended to restrict the children to those ideas which are suggested. It does, however, give support to those children who might be uncertain what they could do. It may also broaden the range from which to select.

Decide on the way that seems to be the most interesting and use it to show the information.

Finally I asked each group to prepare a report on their findings to present to the class.

Children of this age are capable of inventing simple but effective ways of representing their data, especially with modelling experiences.

It would have been unreasonable for me to expect the children to be very precise when making graphs. It was more important that the graphs were clear and organised and helped the children describe and interpret their data.

6 Sorting using rules about two attributes

I introduced this extension of the simple sorting rules activity by explaining that this time I had two sorting rules in mind and they had to figure out both. I carefully chose two rules so that some children would fit both, for example, boys, and those wearing jeans.

The people who fit one rule will stand over near the board and the people who fit the other rule will stand near me.

Eventually students noticed that some people belonged in both groups. I asked them what they wanted to do about the people fitting both rules.

One possible solution is:

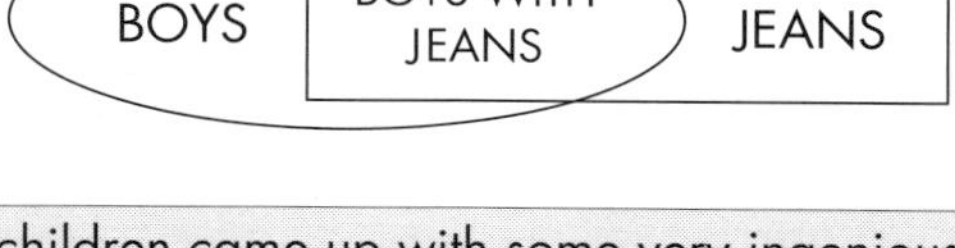

The children came up with some very ingenious ideas. One I remember was 'Nathan and Jane belong to both the groups so today we could put them in this group and tomorrow we could put them in that group. Then it would be fair.' To which another child quipped 'No. Put Nathan in this group and Jane in that group then it would be fairer 'cause it balances!'

I asked the children to describe the groups. They found this quite hard. We eventually developed a description, and I wrote on the board what the children were saying.

This provided a wonderful opportunity for using refining and editing processes which are a feature of our language work.

7 Sorting favourite toys

I began this section by reading to the class the story *My Brown Bear Barney* by Dorothy Butler (Angus & Robertson 1988).

What is your very favourite toy?

As each child made their contribution, I wrote the child's name and the toy on a sheet of paper and placed it on a table where we could all see it.

How could we sort these toys into groups?

We then put the stickers into different groupings as various categories were suggested.

Groups decided on two rules for sorting their toys. They then had to demonstrate this sorting to others, by moving the child/toy names around on the table.

8 Using sorting rules with commercial materials

In my classroom I had a set of Attribute Blocks™. These blocks have the attributes of colour (red, yellow and blue), shape (square, rectangle, circle, hexagon and triangle), size (large and small) and thickness (thick and thin).

I began using a sorting rule to place some of the pieces into a hoop. Firstly I chose just yellow blocks to put into the hoop.

It is important that the children are aware that the pieces which do not 'fit the rule' have something in common. (In this case it is that they are all 'not yellow'.)

When you think you know my rule, help me put the pieces into the hoop.

Why do these pieces go in the hoop?

What then can we say about the pieces that are not in the hoop? Why are they not in the hoop?

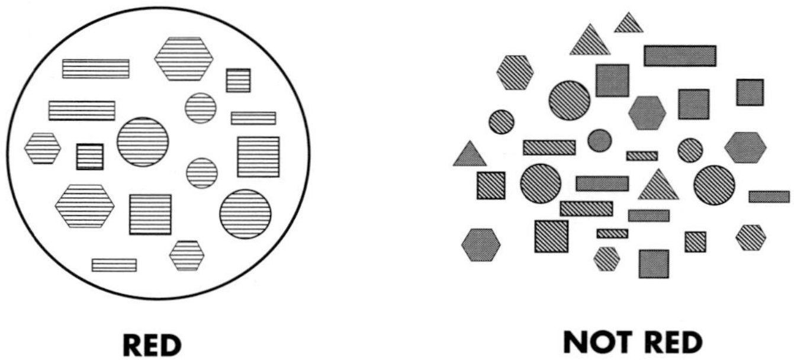

I next placed two hoops on the floor, and started to put blocks into them according to two rules using different attributes. For example, if one is a colour, and the other is a thickness, we could have this.

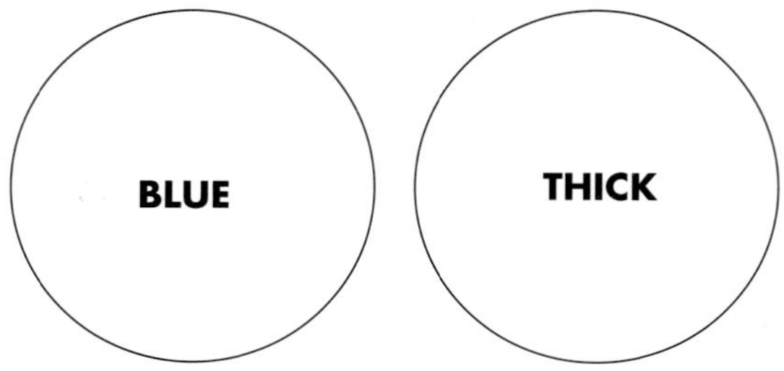

I had to be careful not to try to place any thick blue blocks in the early part of the activity. So all the blocks I happened to put into the blue hoop were also thin, and all the thick blocks happened to be not blue.

Once the children had the rules worked out, I asked them where I could put the thick blue blocks. The children quickly drew on their experiences of sorting people and moved the hoops so that the intersecting sets were clear.

Outside the hoops are all those that are neither blue nor thick.

We tried many of these over the next few weeks until they could comfortably deal with this situation quickly.

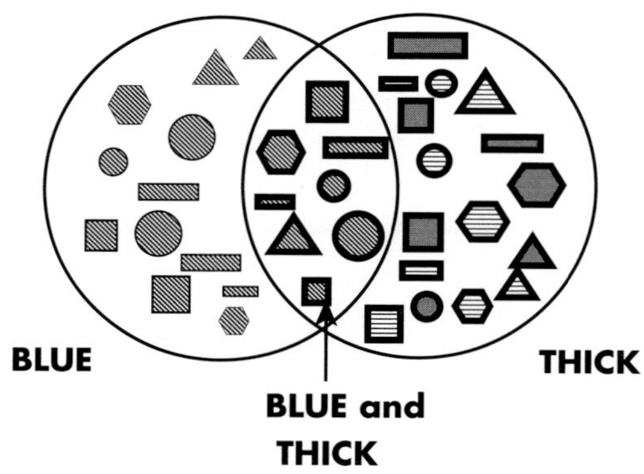

Content epilogue: background information for teachers

Attributes, values and sorting rules

This investigation deals with the ways we place things into categories, according to the things they have in common. We use the term 'attributes' to describe the kind of characteristics that are used for sorting. For example, children have the attribute 'gender'. An attribute that is of use in sorting needs to have clearly defined values, for example, gender has 'boys' and 'girls'. Some attributes such as colour can be very difficult to sort into sets, because of the way one colour merges into another.

In this investigation we have used 'sorting rules' as ways of determining which things (or people) belong to one set and which to another. So the rule RED applied to attribute blocks clearly deals only with the attribute of colour and separates out the RED blocks from all the others, which are then labelled NOT RED.

When this kind of thinking is used with two rules involving different attributes, the result is overlapping sets. Some blocks will naturally belong to two sets at once. (This is also a common situation with people, who belong to several clubs, families, schools, churches, etc. at the same time.)

The use of hoops is a relatively simple way to introduce this work to children. They are often called Venn diagrams, after their inventor. Another way to represent the situation is using rectangles; this way is called Carroll diagrams, after its inventor, the logician Lewis Carroll, best known for writing *Alice in Wonderland*. Carroll diagrams are shown on the opposite page.

Sorting grids and Attribute Blocks™ can offer challenges to children. These grids need to be large enough for several shapes to be placed in a cell. They can be used much more flexibly if the labels are made separately and just placed as and where required.

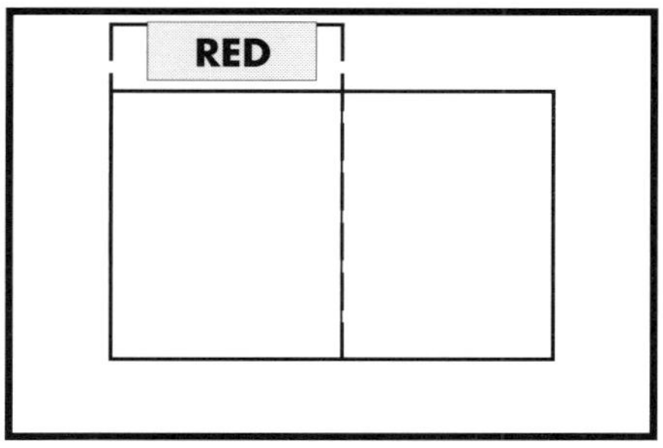

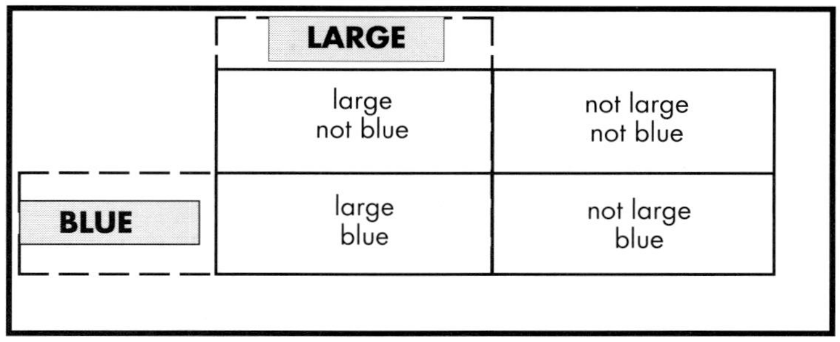

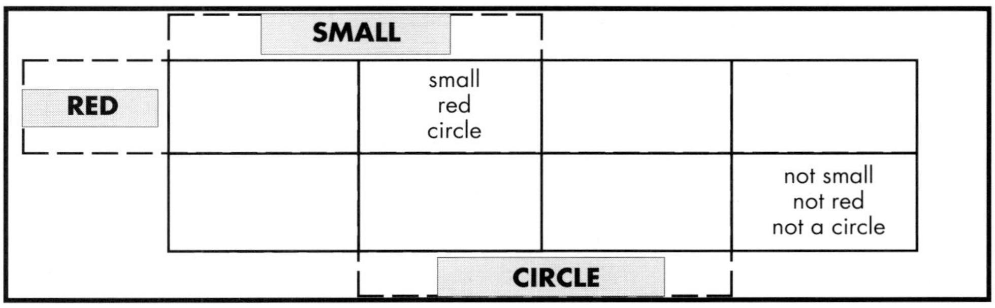

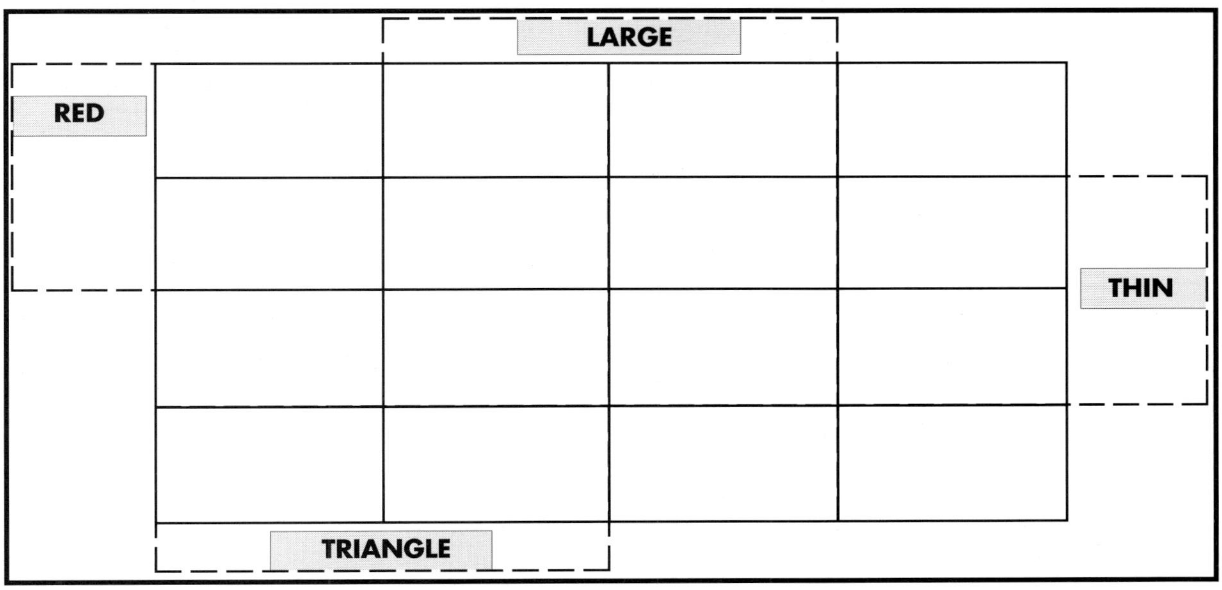

Children can use the grids for sorting things other than Attribute Blocks, for example the numbers 1 to 50 on this grid.

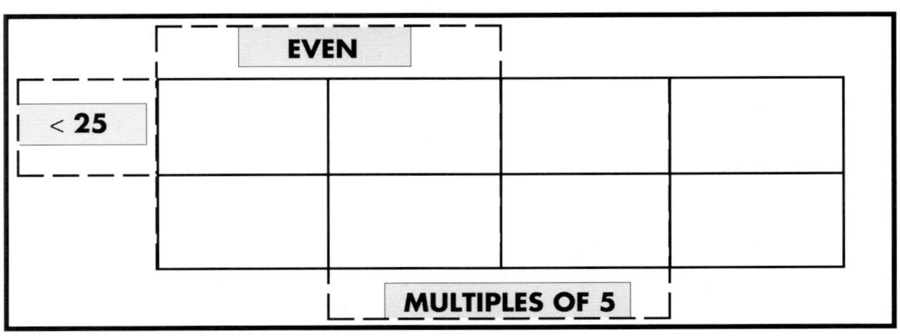

Bikes, monkey bars and skeletons

Years 4 to 8

The three parts of this investigation have a common theme of body lengths, and how the design of equipment based on averages creates problems for big and little children. Statistical techniques are developed gradually: from simple column graphs through frequency graphs to stem-and-leaf plots. *Chairs and tables: just right?* (page 254) is a simpler introduction to the same techniques.

Features

Teaching and learning

- Relates to real and personal equipment used by children
- Uses personal body measurements

Major mathematical content

- Clarifies a mathematical problem from a real situation
- Column graphs and frequency graphs (histograms)
- Stem-and-leaf plots
- Averages (as median values)
- Interpreting graphs
- Comparing two sets of results
- Concept that a large set of results is likely to have a larger spread than a smaller set in the same situation

Mathematical content is further discussed in the *Content epilogue*.

Expected timing

- Each of the three parts should take two to three sessions

Equipment and preparation

- A measuring tape for each small group of students
- A graph grid will help each student
- For *Bikes* you will need access to several different bicycles
- For *Monkey bars* a visit to a set could be useful
- For *Skeletons* you need appropriate art materials (brown paper, tape, paint, etc.)
- Additional data are available on the *Chance and data* computer disk.

A teacher's story

Bikes: They come in several sizes and the seat heights are adjustable, but every day I see that many children do not have the seat at the proper height. We discussed the safe way to sit on a bike, and realised that hip-to-heel distance must match the seat height to bottom of pedal circle. The measurements they then made showed a great range and when they measured their bikes out in the rack they decided to adjust the seat height. This even led to some proposed bike swapping (pending parents' permission).

Monkey bars: The children wanted monkey bars in their own area of playground. I challenged them to find out what height the bars should be so that all children at their year level could use them. They found a way to measure the reach of the children and graphed the results on a frequency graph. I introduced the stem-and-leaf plot. This led to a final suggestion for a sloping set of bars, since one single height would be of limited value. They presented their findings to the school administration, who are still investigating the matter.

Skeletons: How do you make a paper skeleton to best represent the typical person in the class? Working in groups they chose, measured and graphed appropriate lengths, and usually selected the middle value (the median). The skeletons had a life of their own, even parading around the school!

Comments

I had never seen the stem-and-leaf plot before. It offers an interesting way to show the frequency while recording the data. Comparing the histogram and stem-and-leaf plot gave some valuable insight into the thinking processes used.

Generally children were very involved and keen. They were given extra experience of how mathematics affects us all through design.

Summary of the stages in this investigation

1 **Bikes: clarifying the problem**

2 **Obtaining and organising data**

3 **Interpreting the data**

4 **Monkey bars: clarifying the problem**

5 **Obtaining data for our own class**

6 **Frequency graph**

7 **Stem-and-leaf plot: by tens**

8 **Stem-and-leaf plot: by fives**

9 **Extending the data set**

10 **Interpreting the data**

11 **Skeletons: planning an investigation**

12 **Conducting the investigation**

13 **Discussion: what have we learned?**

14 **Consolidation**

1 Bikes: clarifying the problem

One day I rode a tiny BMX bike into the room. After the laughter subsided the children quickly told me that I was too big for the bike! (Or was it that the bike was too small for me?)

The issue of how you tell when a bike fits you (or is it you fit the bike?) needed some discussion. The bicycle education program had definite ideas about that.

How can you tell whether a bike is too big or too small for you?

One teacher had a tall student bring her BMX into the room and ride it around. Another also brought a ten-speed bike in for comparison.

2 Obtaining and organising data

In pairs, students measured their 'hip-to-heel' in centimetres (or they could do it directly with wool).

We displayed the data as a column graph, each person's hip-to-heel measurement showing as one vertical straight line. This led to an appreciation that the range of bike sizes needed was quite large.

The results could range from 70 cm to 100 cm.

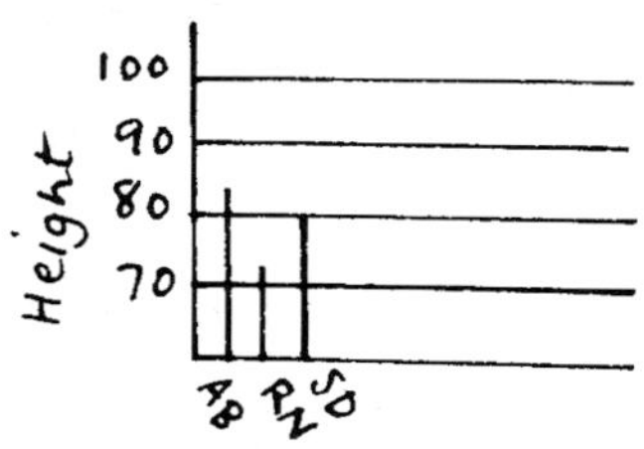

I drew lines horizontally at each 10-centimetre mark so we could easily see how many children had hip-to-heel lengths of between 70 and 79, 80 and 89, 90 to 99.

I then drew a *frequency graph* in which each child is represented, not by a line, but by a rectangle. They put their initials and hip-to-heel length in their own rectangle.

It was important that they could see each individual piece of the data on both forms of graph; going to a frequency table first would have lost that.

Many of the children ride their bikes to school, so those who do were a great help in this lesson. We all went outside and made the required measurements on the bikes as they stood in the bike rack.

The critical length on the bike is from the seat to the bottom of the pedal circle, as shown on the diagram, and this has to be about equal to the distance from hip to heel.

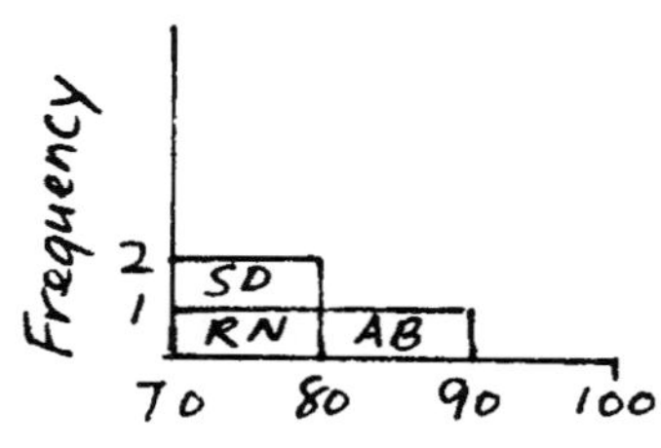

Extra data for 71 children are available on the computer disk in the field called 'hip'.

3 Interpreting the data

For each bicycle, how does the seat to pedal distance compare with hip to heel of its owner?

In some cases the difference between the body measurement and the bicycle measurement showed us that the children had outgrown the adjustment of the bike.

To check the mathematical answer, we asked each child in turn to sit on the bike while others held it. We all judged whether the seat height needed adjusting. With a spanner, supplied by me, most of them made the adjustments until the length was as required. Bikes were then safer, easier to ride and more comfortable.

It is important to check the mathematical prediction of a solution to the problem by a direct test wherever you can.

Some children couldn't accept that their bike seats were too low. It seems that some aspects of safety are not obvious.

In two cases small children had adult bikes, and their riding method forced them to avoid sitting on the seat. This meant they had less control over the bike than they should have, and we were all reminded how important safety is.

In one case a tall child was still using a tiny bike from earlier years. This also led to a reduction in control, and two children proposed a bike swap for a few months. They promised to check with their parents.

4 Monkey bars: clarifying the problem

The next thing they wanted to investigate was the *monkey bars*, their name for the climbing frame in the older children's playground.

The shorter children complained that it was too high, while the tall ones scoffed, saying it was easy to reach. We then went out to the frame so that they could demonstrate the problem to me and to each other.

As there are no monkey bars in the area used mainly by these children, I posed the following challenge.

How high should the bars be for a frame to be used by all the children in our year level?

I asked them in groups to tell me what body measurements were important in the design of climbing frames.

In no time flat (I didn't really prompt them, well not much!) they had realised that they needed to know how high all the children in the year level could reach, with a small jump. They decided that the bars needed to be just a little above the reaching height, so the jumping needed was not too much.

5 Obtaining data for your own class

Organising the jumping height works well this way. I taped a chart on a wall, and let children take a standing jump up the wall. They allowed for the grip on the bars, and each person soon figured out how high they would like the bars to be. It was great fun!

I organised them into groups to get the information. Each group took their own measurements, and wrote them on the board.

> If the jumping complicates matters too much, just measure how far they can reach up the wall.

> Extra data on the computer disk are in the field called 'stretch'.

6 Frequency graph

Again we drew a column graph with horizontal lines at each 10-centimetre mark to show how many children had reaches of between 120 and 129, 130 and 139 and so on.

> One method to graph the outcomes is by drawing lines on a graph grid on the overhead projector, and then photocopying.

We also drew a frequency graph and again they put their initials in their own rectangle.

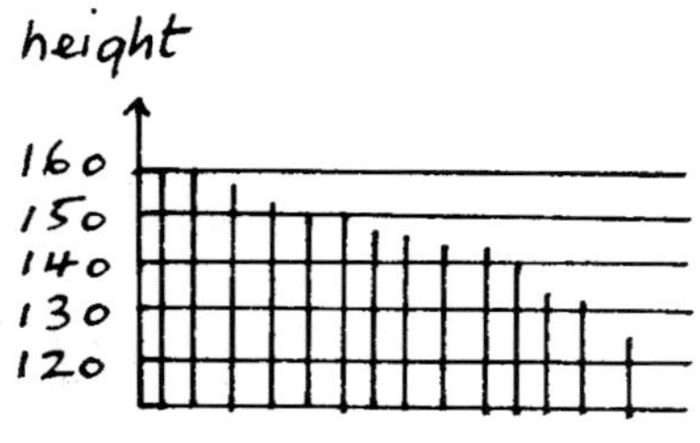

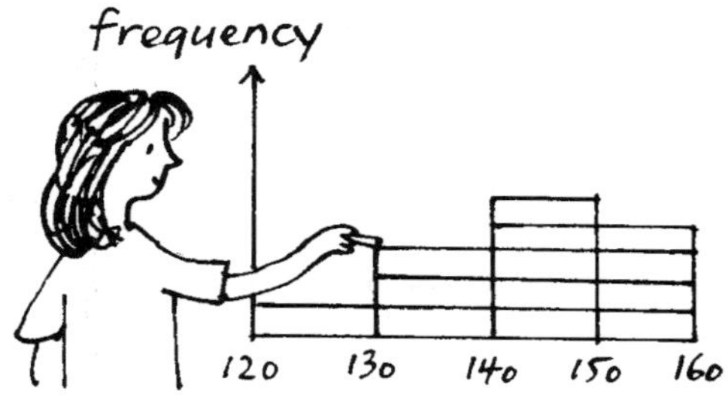

7 Stem-and-leaf plot: by tens

I showed them how to make a stem-and-leaf plot from their data. I set up a table showing the 'stem'. The stem was the number of tens. I showed them how to place the ones, ranging from 0 to 9, in the spaces alongside their stem; for example, 15 | 6 means 156 centimetres.

After I explained how to enter data in the table, I asked them to come out and put the number of units (or ones) on the correct line of the stem table I had drawn on the board. They knew to which part of the stem the units fitted; they just had to look at the number of tens.

The stem-and-leaf plot showed all the data and it was easy to see how it was spread out.

Stem	Leaves
16	0
15	0 6 1 8
14	9 3 5 7 2
13	3 5 7
12	6

> So 150, 156 and 151 fit into the table like:
> 15 | 0 6 1

8 Stem-and-leaf plot: by fives

I also showed them how to draw a stem-and-leaf plot with stems 5 units apart. In the table shown, each ten is split into two parts: the digits 5 to 9 on the upper line, and the digits 0 to 4 on the lower line.

I asked the students to come and put their own value into this table. It gave a more spread-out shape, and told us more about how our numbers were spread.

How would you describe the way the numbers are spread out to another class?

Do they bunch up around one or more values?

Are there gaps? What do the gaps mean?

Are there any exceptional values: unusually big or small reaches?

Stem	Leaves
16	0
15	6 8
15	0 1
14	9 5 7
14	3 2
13	5 7
13	3
12	6

These questions deal with ideas about describing a *distribution* (the way the values are distributed or spread out). It is not easy to describe a distribution in words, but it is the essence of graphical interpretation. Students will need many tries at this; and should evaluate each attempt (in terms of the information that is conveyed).

9 Extending the data set

We decided to get data from other classes at our year level. I asked the groups to suggest how this could be done most efficiently.

Our decision was that I get permission from other teachers for a small group from my class to take the measurements at an agreed time. Each group was asked to propose a method of taking measurements quickly and efficiently. Methods were tested to see which was best. This was easily decided and the procedure was finalised.

This is an optional extension depending on available time and class motivation.

Measuring the 'reaches' of the other children took most of a day, on and off. Finally all data were in, and they realised why the frequency graph was useful; they had so many numbers that drawing a separate line for each was impractical.

How is Ms Majid's class different from us? How is it the same?

What happens if we put all the measurements together?

Each group used a stem-and-leaf plot (by fives) to show the data for the class they had visited. The results were displayed on the board for easy comparison.

They then worked in groups to get all data into a new stem-and-leaf plot.

To achieve this simply, I reorganised the groups so that each new group had one member of each team who visited the different classrooms.

Is the combined stem-and-leaf plot for all the year level classes the *same* shape as the stem-and-leaf plot for our class? How is it *different*?

This led to recognising that the larger population of which we are a part has a greater *spread*, and a slightly different 'middle' value.

10 Interpreting the data

We now came back to the question that motivated all this:

At what height should monkey bars be for our year level?

It was clear that a few centimetres above their reach would be suitable for short children but would be of no use at all to tall children, and vice versa. I put them into groups to discuss a possible solution.

One group suggested two frames, another suggested three and recommended the heights needed. However, the preferred idea was for a long sloping frame, starting at the lowest usable height and gradually rising to just above the greatest usable height. The group showed that most kids could use most of it, but there was some part of it for all, and a challenge for shorter kids to jump as high as possible.

A school has found a manufacturer of these ideal frames! They even have a platform at each end to assist smaller children to get on and off the bars.

11 Skeletons: planning an investigation

This is described in the form of a more open problem for the class. By this stage the students may be able to take much of the initiative. If this is not the case, make suggestions along the lines of the methods developed in the previous sections.

How could we make a pretend skeleton to best represent the size of our class?

This discussion led to the following, all the ideas being raised by the children — I didn't have to drop many hints at all!

- We decided to use paper strips cut from old newspapers.
- Each group was to make their own decision about which parts of the skeleton to make.
- Once they knew what lengths were needed they would get the appropriate information from each other. They decided to measure each other in pairs first and write down lengths on a sketch of themselves.
- Once they had all the lengths they had to draw either a column graph (one line for each person), a stem-and-leaf plot or a frequency graph. They would use this to help choose the best size to show our class.
- They would choose the size for the skeleton, and write a sentence or two to say how they decided.
- They would make tubes or strips of paper the right length, and attach them to the rest of the skeleton with smaller strips of paper.

12 Conducting the investigation

All went fairly smoothly, but I had to help with the stem-and-leaf plots and with the frequency graphs.

Basically, they needed guidance on the spread of the values to be included in each group for the frequency table (or part of the stem). I tried to split them up in the simplest possible way, with three, four, five or six stems or groups. It had to be easy for them to decide into which group (or onto which stem) the values went.

Stem	Leaves
6	0 1 2
5	0 2 6 6
4	0 0 4 7 (9) 9
3	7 8 9 9

median

For most body parts there was one 'middle' length, which mathematicians call the median. Most children thought that this best represented the class, and I agreed. Each group then used these median lengths to cut paper strips of the right length and create their skeleton.

When that was over I had them all write a story about how they had created a typical skeleton for the class, and the creativity began. They gave them names, and invented families for them and all!

The Addams family got plenty of mention at this stage, since the movie was in town, and the series was back on TV.

I arranged a parade of the skeletons around the other classrooms at our level. A spokesperson for the groups had to explain to the other classes what they had done. They did a good job with the explanation.

13 Discussion: what have we learned?

At the end of all this enjoyable activity I wondered whether the students realised how much they had actually learned. To find out, and to help them realise the number of mathematical things they had been doing, I held a discussion, and then asked them to write a report for their parents on what they had learned.

What mathematical things did we do when we were working with bikes?

What did we do when we explored the height of monkey bars?

What mathematical things did we do when we made our skeletons?

In what other situations might you need to use the same mathematical knowledge?

In this discussion, I tried to help them see that they had used a great number of skills that could be applied in other situations. Some were techniques, such as measuring, grouping data, tallying, making a frequency table and drawing a kind of graph. Some were verbal things, such as asking the right questions, trying to understand the meaning of the graphs, explaining things using mathematical language. Some involved personal skills, like group cooperation and problem-solving strategies.

They soon realised that the ideas of measuring and graphing are very useful tools. The range of suggested applications stretched to shopkeepers (market research), designing playground equipment, designing clothing in different sizes, making furniture, etc.

14 Consolidation

I found the worksheet very useful to review the major mathematical ideas and techniques after a short break from this kind of work. It reminded the children of the skills in drawing stem-and-leaf plots from real data, and in interpreting them. It recalled the fact that the plot looks different depending on the numbers on the stem.

> Additional data on child heights are on the computer disk. They could be used for further graphing practice and comparison.

The worksheet also asks them to make a frequency table and graph from the data. They will see that this gives the same information about the way the numbers are distributed as does the stem-and-leaf plot, but loses some detail along the way.

Content epilogue: background information for teachers

Converting column graphs to histograms

This investigation clearly has a lot in common with *Chairs and tables*. Many of the mathematical techniques used here are described in the *Content epilogue* of that investigation.

A column graph is made up of columns with heights that have particular values. Each column can have a label; in this investigation it is the name of the student. The heights of these columns can be spread over a fairly large range. If we split that range into about five or six equal smaller ranges, then we can count the number of columns that belong in each smaller range. (The tops of the columns show which range.)

Then this count (called the frequency) can itself be the height of new columns. This new graph is called a frequency graph, since it shows frequencies. Its special name is *histogram*. There will be five or six columns, one for each of the smaller ranges we used. The frequencies must be whole numbers.

Stem-and-leaf plots

This relatively new method of 'graphing' or 'tabulating' data has two clear advantages. Since you don't simply count frequencies in small ranges, you don't lose the actual numbers. If you can get a feel for the likely range of the data before you start, you can set up a stem-and-leaf plot and record the numbers as they are measured directly on the 'graph'. It helps a lot to make each number exactly the same width, so that the length of each row shows how many numbers are in it.

> The best way to start is to plot by tens, as in the monkey bars investigation. If there are too many in each tens row, then you need to split the tens into either fives, as we have done, or even into twos. (Anticipating this eventuality first can save a lot of work later.)

Stem	Leaves
3	0 4 6 8
2	1 5 8

> It is frequently very useful to put the numbers in order, with 0s closest to the stem, and so on, so that you can see clearly how many there are of each. (This is very handy when you want to find the median value.)

Types of averages: mean, median and mode

The skeleton investigation requires students to decide the typical (or average) length of each body part. The hard way to do it, and not necessarily the best, is the common average, called the *mean*. If this is to be done, have them use a calculator.

The middle value when they are arranged into order is called the *median*. It is often very close to the common average. If you have created a stem-and-leaf plot and arranged the numbers on each stem into order, you have really arranged the whole set of numbers into order. So you can count from the bottom, or the top, until you have found the middle one (or two). If there are ten values the middle is halfway between the fifth and the sixth, and that is the actual value of the median; it will have five values under it and five values over it. The same general idea is used for any even number of results.

The most common result (or several of them) is called the *mode*. It is easy to find, but of course there may be two of them. If this happens you really have to use them both, and the situation is called *bimodal*.

Worksheet

Hip-to-heel lengths and bike size

While investigating bike sizes, a class measured the distance from hip to heel for each student in centimetres. Here are the results:

85	86	91	87	77	88	83	86	74
89	85	85	80	94	82	84	89	84
94	84	76	93	86	84	94	84	89

1 Make two stem-and-leaf plots from the numbers.

By tens

9

8

7

By fives

9

9

8

8

7

7

2 Write a sentence about the hip-to-heel lengths of this class.

3 Which of the two stem-and-leaf plots is more helpful? Why?

4 Complete this frequency table. Use it to draw a frequency graph.

Range of lengths (cm)	Frequency

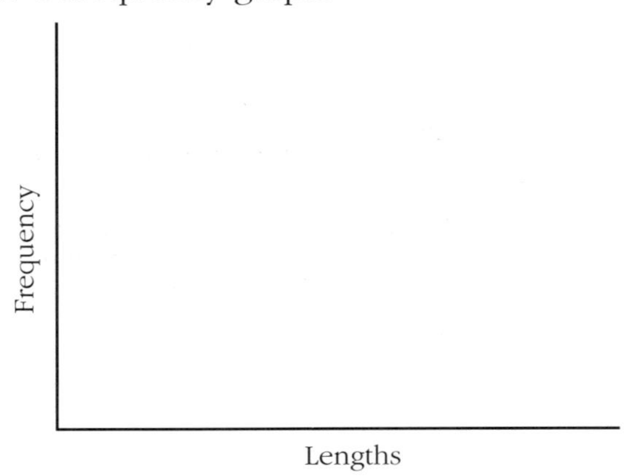

5 Which do you find more helpful — the frequency graph or the stem-and-leaf plot? Why?

Popcorn

Years 4 to 7

Children decide to make popcorn to raise money for charity. But how much popcorn should they make? What kinds? How much should they charge for it? These and other questions form the basis for developing an understanding of how to use sampling techniques to obtain useful information.

Features

Teaching and learning

- Group work
- Realising the role of mathematics in marketing
- Preparing and administering questionnaires

Major mathematical content

- Statistical surveys
- Collecting and interpreting data
- Understanding simple proportional relationships
- Comparing statistical data
- Modelling the population with a sample
- Experiencing sampling methods
- Drawing inferences from data collected from a sample population
- Money and measurement practice

Mathematical content is further discussed in the *Content epilogue.*

Equipment and preparation

- Be prepared to actually make the popcorn (A recipe for honeyed popcorn is found in the *Content epilogue.*)

Expected timing

- The preparation, up to the making and selling of the popcorn, could take about 2 weeks depending on how thorough you choose to be about the market research

A teacher's story

Each class was to raise money for a chosen charity. My Year 5 class wanted to make popcorn and sell it. I felt it was a good opportunity to involve the children in a range of ideas to do with statistics and sampling in the context of a *real* project.

Planning the project involved discussing a number of questions with the children: what types of popcorn would we make? How much should we make? How much of the ingredients should we purchase to make that amount? How much should we charge? The children quickly realised that these questions were interrelated and that it was not really useful to guess at possible answers. By group discussion we set priorities and with some direction from me, carried out a range of sampling procedures so that these questions could be settled and the project successfully undertaken.

Overall the project took about three weeks. During the first week we discussed the project, carried out an initial survey, and made a sample batch of the popcorn to enable quantities and costs to be calculated. During the second week we undertook a more detailed market survey which involved taking into account the effect of different pricing on anticipated sales. The main purchasing of ingredients and final production and sale were carried out in the third week. This was followed by a comparison between the amount of popcorn actually purchased and the sales anticipated by our sampling methods.

Comments

I found that some children needed more work to develop a basic sense of proportional equivalence (for example, that 5 out of 20 has a relationship to 25 out of 100). While some children are able to use equivalent fractions or percentages to express these relationships, other children may need help within the investigation to gain a real sense of this.

Summary of the stages in this investigation

1 Planning the project: the need for a survey

2 Samples representing the school population

3 Choosing sampling methods

4 Choosing questions for an initial survey

5 Running the initial survey

6 Thinking about prices and costs

7 Finding production costs

8 Market research survey

9 Deciding selling prices

10 The popcorn sale

11 Comparing sales with sampling estimates

12 Report writing

1 Planning the project: the need for a survey

I brought in samples of the types of popcorn we could make and sell so the children could taste them. This stimulated interest and enthusiasm.

I wanted the children to realise that asking everyone would be very time consuming and still not really give an answer as to how much of which type people will actually buy when the time comes. Those who answer surveys don't always tell the truth and may change their minds or forget their money on the day.

How could we find out how many people would buy our popcorn?

We can't say how many people will buy it.

So we can't know how much to cook.

Ask everyone in the school what they're going to buy.

This was designed to get students thinking about the concept of 'market research' to establish a framework for some of these unknown questions.

2 Samples representing the school population

The idea of asking a representative sample of the school population was introduced. I deliberately, but without making it obvious, chose seven girls and one boy, quite unrepresentative!

I wrote a few questions from the class on the board. First I asked the sample and wrote their responses as a fraction, such as $\frac{6}{8}$ expressed as 'six children out of the sample of eight'.

Let's see if asking questions of a small group of children can tell us anything about a bigger group.

What questions could we ask our sample group?

Who rides their bikes to school?

Who likes KF Chicken?

Who likes Coke?

So if six children out of the sample of eight children likes KF Chicken,... and these eight children are the same as the rest of the children in our class,... how many children in our whole class could we expect to like KF Chicken?

It happened that there were 32 children in my class, so eight was a nice neat number to use as a sample which allowed conceptually simple proportional comparisons to be made. You will need to adapt the sample to your own class size if you use this way of establishing the idea.

They realised that the number of children in the class was four times the number in the sample, so that the number of children in the sample (six) who liked KF Chicken could be multiplied by four to get the expected number of KFC fans in the whole class (24).

We multiplied the number of children in the sample who *didn't* like KFC (two) by four and found that the numbers 'added up' (8 + 24 = 32).

We did a quick hands up check of the whole class and found that the number was very close to our estimate.

Tabulating our information

We completed a table on the board to show the following information:

Question	Who likes KFC?	Who likes Coke?	Bikes to school?
Sample proportion	$\frac{6}{8}$	$\frac{2}{8}$	$\frac{4}{8}$
Calculation	4×6	4×2	4×4
Estimated number in whole class	24	8	16
Actual number in whole class	25	20	21

Once we saw that the second two results were not very good estimates of the actual whole class results I asked the children why this might be so. Several children noticed that it was mainly boys who liked coke when we asked the whole class, but that the sample was mostly girls.

> As well as the biased sample I also selected the questions carefully so that one might be gender related; I knew from previous surveys that Coke seems to be more popular with boys than girls.

Children were then asked about other ways we could choose a sample to be a fairer representation of the class for the questions we asked. There were about half boys and half girls in our class so we changed our sample to half boys and half girls.

When we went through the same process, we obtained a closer match between the responses of the sample and the class population.

> Developing the idea of proportional comparisons is central to this stage of the investigation. Unless students see that there is a sensible relationship between a sample population and a whole population which can be mathematically calculated, the idea of comparing results obtained by different sampling methods to obtain the most useful results is entirely lost.
>
> Younger children are likely to be able to understand a relationship expressed as 'nearly half of the children'. But most would need help to see that 14 out of 32 children is nearly half. Expressing this relationship as $\frac{7}{16}$ or 47% of the whole school would not convey the same idea to them.
>
> It's a good idea to expose children to relationships between mathematical ideas long before they are expected to manipulate them. This provides the opportunity to build up the big ideas of mathematics slowly. They need to move easily from one form to another, rather than treating numbers such as fractions, percentages and decimals as quite separate entities.

3 Choosing sampling methods

Children brainstormed sampling techniques we could use to estimate the popcorn sales without having to survey the whole school.

After discussion I listed a range of techniques and arrived at four different sampling methods. These were assigned to four different 'market research companies'; the 'manager' of each being the child who suggested the particular method. These managers then 'employed' their research teams.

How could we choose a sample of children to ask about popcorn?

Ask one class and 'times it' by how many classes.

Older students might have a different preference for popcorn than young children, so you could ask people from each class and 'times it' to make the right number.

The companies and sampling method were:

- Eagles Market Research Co. — ask one whole class.
- Ninja Marketing Co. — ask the three children sitting closest to the door from each class.
- Super Popcorn Sales — use alphabetical order. Ask the child whose surname is first, last and exactly in the middle of the list for each class.
- Matthew's Marketing Company — choose from a hat. Make 32 slips (maximum class size) and mark three of the slips to show who will be asked from each class.

Because of the numbers in our school (about 300 students in ten classes) a sample size equivalent to one class (approximately 30 children) was the obvious choice, so that multiplying by ten provided the population estimates from the sample results.

It was important to choose a similar sample size for each group to work with so that the expected numbers of consumers in the whole school would be comparable. The difficulty level of the mathematics needed for the population estimates must also be taken into account.

The aim was to have a conceptually easy proportional relationship between sample size and total population. We sampled a tenth of the school so we predicted the whole school results by multiplying by ten.

4 Choosing questions for an initial survey

The questions to be asked were selected by the whole class so that the overall sampling methods could be compared. The children quickly decided we should ask how many packets of each type each child in the sample will be likely to buy.

At this stage I purposely did not point out that the cost and size of bag might influence the consumer. The children had not thought of these aspects and I preferred them to carry out this simple survey first to establish a better understanding of the sampling process before introducing any complexity.

Each 'marketing company' organised themselves to carry out an initial survey with their chosen sampling method using the straight forward questions suggested by the class. To maintain fairness I suggested we duplicate the questions we would be asking so that the children in the sample could just write in the number of bags they intended to buy.

Popcorn market survey

Our room will soon be selling delicious popcorn to raise money for the Save the Children Fund. You can help us decide how much popcorn we need to make by answering some questions. Please write down how many bags of each type of popcorn you think you will buy.

Honeyed popcorn .... bags

Salted buttered popcorn .... bags

Plain unsalted popcorn .... bags

5 Running the initial survey

I duplicated the survey sheets the children had made and helped arrange access to classes. I made sure each group understood and carried out their sampling method correctly as planned and also obtained a list of the children in each sample for future reference.

The results from each group were tallied and displayed on a wall chart and the figures were converted to population estimates.

The results were left on the wall for a day so that the children could digest the information and discuss it among themselves informally before a full class discussion of these initial results was carried out and issues affecting the planning of our project addressed.

Not many children in any of the estimates chose the 'salted buttered popcorn' and the class decided that it was not worth the effort of making it and we'd only sell the two more popular types.

Some children in the survey had asked the marketing teams what the 'honeyed popcorn' was and it was agreed we should have had some samples of the popcorn to let them taste it (as I had done with our class when I introduced the project).

Children were asked to consider possible reasons for discrepancies between the results obtained by different sampling methods.

6 Thinking about prices and costs

At this stage I suggested the idea that we can charge whatever we like for our popcorn.

I think we ought to charge $20 a bag and really clean up!

Yeah, we could make heaps of money. If a hundred people bought a bag that would be... stacks!

If we only charged 20 cents a bag, lots more people would buy it. We could make our fortune that way.

The children realised that we couldn't expect people to buy our popcorn at ridiculous prices.

The question of costs and profit margins was raised. We might sell lots of popcorn at such 'give away' prices, we could end up working for nothing and actually have to pay out more money for the ingredients than we'd get back.

The children were able to see that we needed to find out how much the popcorn costs to make so we can set a selling price to cover costs and still make a reasonable profit to donate to our charity. They could also see that the success of the project depended on people buying all the popcorn we made, which emphasised the importance of obtaining an *accurate estimate of sales which also took the price into account*.

The aim was to develop a practical understanding of the complex relationships involved in such a project and the way that changes in one area may influence another. I've found that presentation within a familiar context and simplification of the mathematics involved, allows children to develop a basic understanding of the issues and the way mathematics can help this type of decision making.

7 Finding production costs

At this point (Friday of the first week) we made a trial batch of each type of popcorn. We kept exact records of the amount and cost of ingredients and decided on a standard-sized bag to use. The number of bags obtained from the batch was used to calculate a 'production cost per bag' for each type of popcorn.

The number of bags produced with a certain amount of ingredients was recorded to help make a shopping list for the final production. We also considered how long the batch took to make to help plan the equipment and time we would need.

We talked about the problems of calculating labour costs. We decided to all 'donate' our labour to a good cause in this case!

This part of the project provided much opportunity for the development of important number and measurement concepts.

How we might allow for the cost of the trial batch itself in the production cost per bag when we didn't know how many bags we would finally produce was an interesting and complex sideline to this discussion.

8 Market research survey

We finally arrived at a production cost of about 20 cents per bag for the plain popcorn and 80 cents for honeyed.

When we know how much it costs to make, how do we decide on a final selling price?

How much money do we want to get?

Different numbers of people will buy it depending on how much we charge.

Don't forget how much it costs to make.

At this point I suggested the 'market research companies' carry out a further survey of their survey group, asking how many bags they would buy at different prices. It was suggested to provide a taste from our trial batch and to show the bag size.

Eventually we agreed on three prices for each type ($1, $2 or $3 for the honeyed popcorn and 50 cents, 80 cents or $1 for the plain popcorn), and to ask how many bags people would buy at these prices.

Each 'company' was provided with survey sheets like those used in the first survey but with space to record the expected sales at each possible price, a bag of each type of popcorn to show the amount for the price, and another bag for offering tastes to the children surveyed. The marketing companies were to interview each child in their sample group and produce a written market report for presentation to the class so that we could vote on the survey results we would actually use to determine our price.

Two days were allowed for completion of the survey (normal class work continued) with some class time given on a third day for producing graphs, written reports and charts for presentation.

The 'presentation' was given by each group after each class break, and afforded suitable fanfare with the rest of the class being invited to ask questions and discuss the results. The written reports and charts were displayed on the class noticeboard and voting for the most convincing marketing report was carried out the following day.

9 Deciding selling prices

The final selling prices were decided upon on the basis of the survey results produced by the most convincing company.

The children needed help from me to look at the different relationships between the amounts which would likely be sold at each of the prices, and the expected profit margin in each circumstance. To find the expected profit from sales of each type of popcorn, we multiply the number of bags by the difference between the selling price and the production cost.

It turned out that we would expect to sell much less honeyed popcorn at $3 per bag than at $2 per bag. So the actual profit at $3 per bag would be less than that at $2 per bag. This result surprised many children who expected that the higher the price the more money we'd get; it showed me that they had not really absorbed the idea during our previous discussions.

10 The popcorn sale

Over the next few days we went into sales mode. The popcorn was produced and sold. Detailed records of sales were kept so that the total number of bags of plain popcorn and honeyed popcorn sold was kept and recorded.

I tried to give a *real life* feel by the appointment of various managers who then selected their staff for each phase in a similar manner to the market research companies. An advertising manager 'employed' staff to advertise the sale, a product development manager organised the trial batches and a purchasing manager bought the ingredients (with the help of mum) after obtaining an advance from finance to cover the cost.

These aspects are not described in any detail, since they are outside the strand of chance and data, and they could be varied a great deal while still providing a background purpose for the sampling techniques central to this project.

A production manager oversaw the final production and packaging and a finance manager kept track of the money spent and received (even applying to the principal for a loan to cover the cost of ingredients and bags, which then needed to be repaid with interest). A sales manager and her staff sold the popcorn and kept a record of the sales. All these managers reported back to the class at the end of the project, saying what they had done and how they could have improved their performance.

11 Comparing sales with sampling estimates

The sales manager was in charge of the team actually selling and was responsible for keeping accurate records of the sales of each type of popcorn for each room. Staff sales were also recorded separately as well as any bags of popcorn remaining unsold. Left over ingredients were recorded by the production manager.

The comparison between sampling methods and actual sales was made with the help of a table.

The sales manager provided a figure for the number of extra honeyed bags that could have sold if we'd had enough stock, and we all agreed it was fair to take this number into account when choosing the winning market research company.

		Type of popcorn	
		Honeyed	Plain
Ninja Marketing Co. (closest to door)	Estimate 1 Estimate 2		
Super Popcorn Sales (alphabetical order)	Estimate 1 Estimate 2		
Matthew's Marketing Company (from the hat)	Estimate 1 Estimate 2		
Eagles Market Research Co. (whole class Year 6)	Estimate 1 Estimate 2		
	Actual sales		

12 Report writing

The children had derived a lot of mathematical benefit from this extensive activity, but I wasn't going to let them off without giving a full report of what they had done, with as much documentary evidence as they could still find. They each wrote a report of the activities of their market research company. This included all the decisions they had made, and the reasons they made them.

Also displayed were statements about the sale which could have accounted for some of the discrepancies. For example: the samples were not really representative (gender and year level imbalance) or that the sample did not give 'true' answers. We discussed the issue of chance differences and one child suggested we could try taking bigger samples out of a population to give a better estimate.

The younger children didn't have as much money to spend as the older children. In fact the winning company had not taken the little ones' estimates at face value, but had made a guess at the likely purchases of some of these children, rather than accept obviously abnormal responses. Even though this group thought they might have been 'cheating' by doing this, in the context of the purpose and outcome of the survey, it was an unexpected but valuable demonstration that mathematical common-sense is an essential ingredient in problem solving for real purposes.

Content epilogue: background information for teachers

Honeyed popcorn

You need: 100 g popping corn
100 g butter
$\frac{1}{4}$ cup sugar
2 tablespoons oil
2 tablespoons honey

Method: In a large frypan, heat oil until very hot. Add popping corn, place the lid on the pan, and shake the pan continually until the popping ceases. Place the popped corn in a large bowl.

Combine honey, sugar and butter in a saucepan, slowly bring to boil, and boil for five minutes. Pour the syrup over the popped corn, and mix well until the corn is all coated. Pour quickly onto well greased oven trays, and cool. Enjoy it.

Sampling and proportion

The chance and data content of this investigation is largely based on developing early ideas of sampling. The basic ideas are these:

- A sample is a small set from a larger group.
- A good sample is one which is representative of the total group — it will respond just like the group.
- We ask questions of the sample, and get the fractions of the sample that respond in certain ways.
- We use these sample fractions as estimates of the fractions of the total population that will respond in those ways.
- The estimate of the numbers in the population can be obtained by using simple proportion (e.g. if the sample is one tenth of the population, multiply the number in the sample by ten) or by getting the same fraction of the total population.
- The sample may not be fully 'representative' so the final estimates may differ somewhat from the 'perfect' estimates; this is not purely a matter of chance, it depends greatly on the choice of the original sample.

At this stage we are not interested in discussing the techniques for choosing random samples, and the many complex ideas (such as confidence intervals) that come from their use.

The ideas that many children will find difficult are those related to equivalent fractions or proportion. These ideas take many years to develop to the formal stage, but their development is dependent on activities like this one to stimulate thinking and create the links between many different ideas that come together in the concept we call 'proportion'.

6 Using secondhand data

Secondhand data is data provided by someone else. In this way it differs from the personal data used in chapter 5. Secondhand data can come in many forms: lists, tables, graphs and even videos. It can also come in the form of databases (computer programs with data in an accessible form), and these are the subject of chapter 7.

The five investigations in this section have very little else in common but their provision of secondhand data. The data are in the form of lists, tables, graphs (many forms) and video film. The emphasis is always on correct interpretation of the data, but in some cases students will need to exercise caution not to be tricked by the presentation.

Two of these investigations (*Dying for a smoke* and *Sunsmart?*) are based on statistics of cancer incidence and death, and are concerned with a serious social issue. *Top of the class* deals with an important issue for those interested in tertiary selection or possible judging of sports competitions: how best to rank candidates on the basis of several scores. *Population centre of Australia* uses geographic and population data to explore a concept of interest to demographers, while also teaching weighted averages. All of these provide data in fairly conventional forms.

Video data offers film of events that can be replayed and paused, and from which classification, timing, counting, reading codes, etc. are all possible. Some of these are obtained from film of live action, and some from radar pictures of violent storms. Questions to ask are offered as suggestions, but the challenge is really about problem posing.

From *A National Statement on Mathematics for Australian Schools*

(Overview) The achievement of confidence and competence to make sense of and interpret data which have been collected, organised, summarised and represented by others is a major goal for this strand. If they are not to be subjected to the kind of exploitation implied by the expression 'you can prove anything with statistics', students need experience in judging the quality and appropriateness of data collection and presentation for answering the questions at hand. They should interpret tables, charts and graphs which range from simple to complex, and from those which provide very good models of data presentation, through those which are flawed but still communicate in useful ways, to those that are misleading.

(Band B)...children should begin to locate and interpret data from a variety of sources, including encyclopedias, textbooks, newspapers and magazines, and to incorporate the results of their research into reports of their own.

(Band C) Many interpretations of data are based on summary statistics, such as measures of central tendency, variability and association, and graphs such as line plots, histograms, stem-and-leaf plots, box plots, scatter plots and lines of best fit. Students should be able to interpret these various representations, understand the conditions under which their use is appropriate, and compare and select from different possible representations of the same data. (pp. 163, 168, 173)

Video data

Years 4 to 12

Not all data is numbers, much can be presented visually. From sports games, to cars on a freeway, to athletes wired up to measure their fitness, to the pace length of the winner of the Melbourne Cup — in all these activities, the data that pupils need to solve problems are available by measuring, counting or timing directly from the visual footage.

Features

Teaching and learning

- Visual presentation
- Group work
- Problem posing and problem solving

Major mathematical content

- Through the snippets, and the problems pupils choose to set and the ways they choose to solve them, a large range of mathematical topics and skills are called into play

Preparation and equipment

- Check the operation of the video and preview the appropriate snippet

Expected timing

- Variable, according to depth of treatment (*Transport data* took the longest in trials, up to several sessions, while *Screen saver* tended to be the shortest.)

A teacher's story

I liked the interactive way that the MCTP snippets brought real situations into the classroom, and that the information students needed is contained in the video. Pupils have to work for it by measuring or counting or timing directly from the screen. These snippets are very similar in style.

I also like the openness of each situation. The video data is an unstructured but rich collection of data about which many questions can be posed. These can then be solved at different levels of sophistication.

The small collection available here covers a variety of situations, from the rather abstract screen-saver program to issues of public transport in the freeway data. I tended to use each vignette to supplement other classroom work.

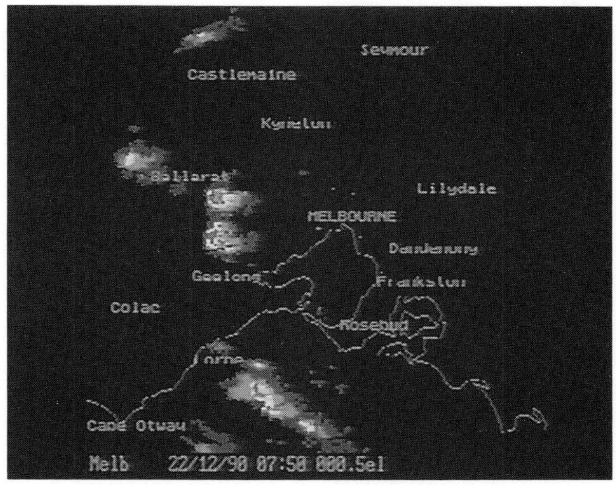

Summary of the activities in this collection

1 **Pace length**

2 **Screen saver**

3 **Transport data**

4 **Weather storms**

5 **Fitness data**

6 **Tennis**

1 Pace length

Two snippets are available, horse races and dogs. The horse race is the 1990 Melbourne Cup won by Kingston Rule. The four dogs are family pets at an obedience club which walk, jog and run over a 20-metre track. The opportunities for problem posing and solving are considerable, and can cover several interrelated aspects of mathematics.

Racehorse pace length

Estimation segment
I conjured up a visual image for pupils of a racehorse and the likely pace length. Pupils each marked out in the room their estimate of the length of one stride.

A racehorse in full stride has a pace length of around 7 to 9 metres, a result which pupils found very hard to believe.

Problem solving segment
I then showed the race and we discussed how we could calculate the pace length from the tape. They noticed the 200 metre post (the one 'just above the tower' on the tape) which we used to count the paces over the last 200 metres.

Counting and calculating
We had fun trying to count the paces in real time, before counting them in slow motion. A calculator helped with the division. We then marked out the result for the winner, Kingston Rule, across the room. Pupils were astonished by the size of this result.

The result was even more surprising when we considered Kingston Rule had just run 3200 metres with a jockey on its back. A fresh horse in a sprint race would have an even longer stride. We determined to check this using the sports news of the races on TV.

Dog pace length

The four dogs are Dyna, a Silky Terrier, Sean, a Corgi, Lara, a Rottweiler and Dennis, a Great Dane. During the introductory scenes, pupils had fun recognising the breeds and comparing with those they owned.

This led us to a separate survey of all the dog breeds owned by pupils in our school. Pupils estimated what the most popular breeds would be — all were surprised by the high percentage of families who had a dog.

Some sample investigative questions are:

1 Estimate and calculate which dogs will have the longest and shortest pace length when walking.

2 Estimate and then calculate which dogs will have the longest and shortest pace length when jogging and running. Enter your results into a table.

Because the number of paces to cover 20 metres was never exactly a whole number, there was much valuable estimation of fractions. A calculator was a natural and useful tool to help with the divisions.

	Walk	Jog	Run
Silky Terrier (Dyna) Corgi (Sean) Rottweiler (Lara) Great Dane (Dennis)			

3 Other investigations are also possible, such as:
- Pupils can obtain similar data for their own pets.
- Which dog has the greatest increase from running to jogging?
- Would percentage increase be a fairer method of comparison?
- Graph the height to pace-length ratios.
- How does pace length compare as a ratio to height of the dog?
- How does pace length in walking compare between humans and dogs?
- You can prove anything with statistics. (I divided the class into four groups and allotted each a dog. Their challenge was to try to prove statistically that their dog had the longest relative stride.)

The height comparisons will need to be estimated (or measured) from screen. For example, does a dog twice as high have a pace twice as long?

2 Screen saver

I used to watch the screen-saver balls moving round my computer screen and started wondering if they conformed to the laws of probability. Putting an electronic timer on the screen and lines to separate the screen into quarters allowed me to ask my pupils the same question.

It is really just the same as a typical textbook question, but the data is visual and the answer is available by timing the events on the screen.

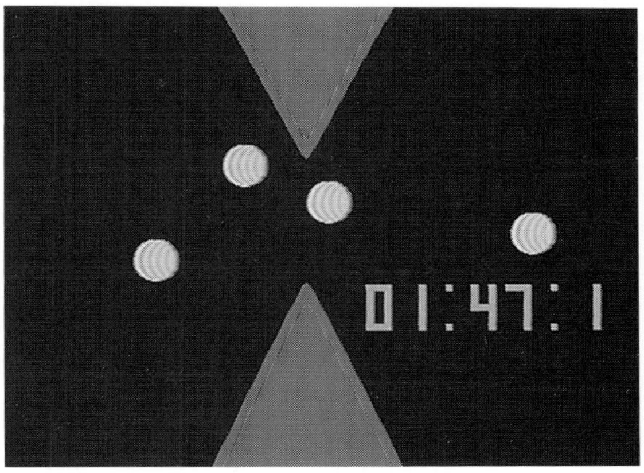

There are several segments on the tape, either with plain background or with colour shading. After posing questions directly, allow pupils to invent questions for themselves. Suggested questions are:

Questions could range from very simple to very complex.

If each ball moves randomly, then it should spend equal times in each half of the screen. Does it?

What is the probability of a collision between two balls?

What fraction of balls should 'get through the gate'?

My senior students got really involved, arguing how a collision should be defined, and then collecting the data to support (or deny) their theories.

We decided to use 2 minutes of action. Each time the ball moved from the top half to the bottom half, we used the pause button to note the time.

For two balls, what percentage of the time should they both be in the same half? Repeat for three and for four balls.

What is the probability of a ball passing across the centre spot (indicated by all four colours) within 1 minute of action?

The pupils appreciated that the answer was available from the screen data rather than looking up the 'back of the book'.

3 Transport data

It seems so simple to provide raw data showing traffic flows along a roadway, but as a data source it is extremely rich, allowing many different questions to be posed and then answered using data from the screen.

There are three distinct segments on the tape and each is appropriate for posing a number of different problems.

I started with some structured questions, but as pupils became familiar with the tape, they began to pose the questions. Some suggested questions follow.

What are the most popular colours?

The class all recorded their estimates, revealing a fascinating variety of perceptions, and stimulating very close scrutiny of the tape (and much 'discussion' about the definition of colours).

One of the pupils brought in a video movie from another country that showed freeway data. We hypothesised and tested whether there was any difference between the two countries. There was!

How long does it take to see a million dollars pass by?

This problem caused much enjoyable discussion, and the answer was most unexpected. The times guessed by pupils for the segment we chose ranged from about 30 seconds to 10 minutes. After researching the approximate price of certain cars and making allowances for age, the pupils were stunned when the answer turned out to be much less.

What is the average age of cars on the road?

Using number plates and their known release dates (by starting letter), the approximate age of each car can be estimated. The following table is from the Victorian registration figures.

We sampled 50 cars, each small group doing four or five, accumulating the results on the board in a stem-and-leaf display, and calculating the average.

Number plate starting letter	G	H	J	K	L	I	A	B	C	D	E
Release date	1953	11/59	12/64	10/68	11/71	10/74	11/78	12/81	11/84	7/87	12/89

> In our State we hypothesised that the result would be different from the Victorian result. We obtained the release dates from the motor vehicles registration department and worked out a sample from local traffic.

What is the average number of passengers per car?

> This data and the result led us into comparisons between public and private transport. For the 100 cars we sampled, assuming they all drove an average of 20 km on their journey, how much did it cost in petrol? What would it have cost by train? Or bus?

What is the capacity of the freeway?

At this density, how many cars pass the counting point in 1 hour? How many people could pass in 1 hour? How many people actually pass? Relate these ideas to the need for freeways to get part of the work force to work on time.

What are the most common types of vehicle on the road?

Pupils estimate the relative percentages (sedans, utilities, trucks, buses, four-wheel drives) and then check from the data.

What is the average speed of vehicles?

The pupils loved the 'amphometer' concept of the electronic lines on the road (which are 100 m apart) and the clock on the screen which allowed them to work out the speeds of the cars.

Estimate the probability of a particular number plate:
- **having multiple digits**
- **ending in an even number**
- **having all odd digits**
- **having digits in ascending order.**

> I wrote several questions on the board — each group chose a different one — they all did their calculations and wrote their answers on the board. Then, as a group, we checked each one from a 100-car sample. They certainly counted with some intensity, hoping their calculations would be validated.

Number-plate cricket

We played number-plate cricket — a zero meant a wicket fell, all other digits scoring runs. Two halves of the class each represented teams. You should have heard the groans and cheers when a 000 car appeared.

4 Weather storms

Four rather violent weather storms are captured on radar. Using the intensity of rain (shown by colour) and the known area of the terrain (from the background map), many questions can be posed and solved.

Pupils will recognise these types of vignettes from the weather reports they see on TV. These 'movies' are actually made from a series of stills (showing data 10 minutes apart) which are then displayed at high speed.

The four segments are:

- a severe thunderstorm passing across Melbourne (22 December 1990)
- a severe thunderstorm passing across Sydney (20 January 1990)
- a cyclone across the coast of Western Australia (23 April 1990)
- a hurricane passing across Darwin

Each segment is shown twice, the second one slower than the first.

Storm data on radar: background information

The Australian Bureau of Meteorology provides a radar alert for the detection of violent storms and cyclones through the use of satellite pictures and a large number of radar transmitters located around the coastline. In addition to this 'on-the-spot' information-gathering system, a number of computer simulations are used to predict what could be expected to develop in common situations. Warnings are sent to police and State Emergency Services, and to the media.

RAPIC is the name of the system used by the Australian Bureau of Meteorology to observe rain and hail. The coloured images show the intensity of the radar reflections. In general the more rain the more reflections, so the colours tend to show the heaviness of the rain. The method of displaying the radar information used on the videos is the same as that used on TV stations, since TV channels use the same data by arrangement with the Bureau of Meteorology. The colour code is shown on the worksheet. The images are 10 minutes apart.

The Melbourne storm

On Saturday 22 December 1990, an intense cold front developed in the Bight. A single storm started in central Victoria at about 6.30 p.m. (1830 hours). It then split into two storms, which is a common phenomenon, one travelling over Geelong and the other over the Melbourne area. The Melbourne storm caused the most damage, as it moved at about 60 km/h to the east. It kept up its intensity for at least 1 hour, passing Footscray at 7.30 p.m. and Balwyn at 8.05 p.m.

It produced widespread hail damage throughout much of the Melbourne area. The wind damage was due to two short-lived tornadoes which rotated clockwise as they passed over the Footscray and Balwyn areas. The roofs of a motel, a cinema and a school were removed, and trees were ripped out. Two railway cranes (of 229 tonnes) were moved 100 metres to the east and derailed at the end of the track. In each case the size and pattern of the damage suggested that a tornado rotating clockwise was involved. Wind speeds reached about 120 km/h in the northerly edge of the tornado, while the southern edge would have experienced relative calm.

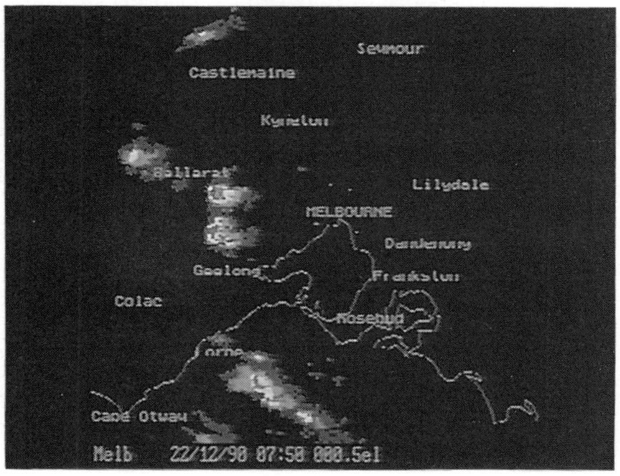

The Sydney storm

On Saturday 20 January 1990, a severe storm crossed Sydney airport (Mascot). This involved lightning, thunder and hail, as well as strong winds. The violent changes of wind direction that are involved as such a storm passes are shown on a wind speed and direction chart used by the Bureau of Meteorology. This chart shows that, at about 5.30 p.m., the wind moved from 20 km/h gusts from the north-east (about bearing 045), to zero wind speed, and then swung around to 50 km/h gusts from the south (bearing 180).

Cyclone Orson in Western Australia

Cyclone Orson was the most severe of the very many cyclones to strike this part of Australia since records have been kept. The dramatic end to the picture sequence illustrates its violence well.

Orson was monitored by Dampier radar from 3.30 p.m. on 23 April 1989 until about 1.50 a.m. the following morning, *when the radar transmitter was destroyed by the winds,* resulting in a sudden absence of further data. It is this time period that is shown on the video.

During the time it spent approaching the coast it moved almost due south at an average speed of 24 km/h. Although it is not shown on the radar images, the average speed of the cyclone in the 24 hours after it crossed the coast was 43 km/h.

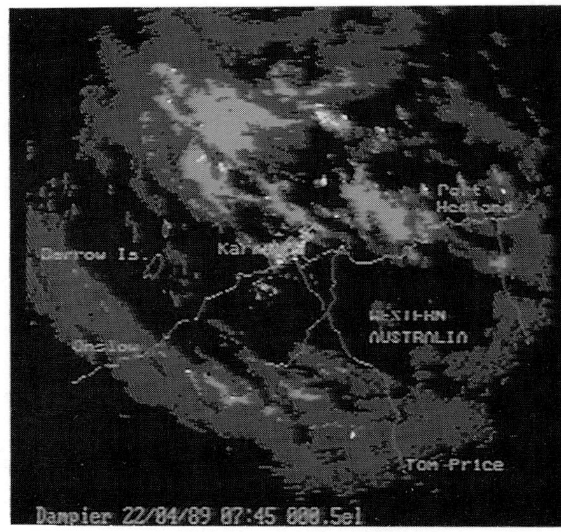

Winds

The cyclone passed directly over the North Rankin A gas platform. As it approached, the wind speeds rose from about 80 km/h to a maximum gust of 275 km/h. The eye of the storm then passed directly over the platform, so that the wind speed dropped to under 15 km/h for a period of about 10 minutes. As the other side of the eye hit the platform the wind speed rose so suddenly that the wind-speed meters were destroyed as the wind reached about 160 km/h. Maximum wave heights during the peak wind period at the gas platform were in excess of 20 metres.

Barometric pressure

The nature of a cyclone in the southern hemisphere is that the winds rotate clockwise around an intense low pressure eye. In the case of Orson, the eye appeared to be about 40 km in diameter. As the eye approached the gas platform, the pressure fell from about 1000 hPa to 960 hPa and then to 905 hPa in about 1 hour. It remained below 910 hPa for about 1 hour, before rapidly rising again as the storm passed on.

Rainfall

The rainfall was only recorded on the coast, and largely because Orson was moving so fast compared to many cyclones, was relatively light for this region. Most of the rain fell on the eastern side of the eye, which is understandable given the clockwise movement of the wind, bringing moisture from the sea.

The storm surge

In such a storm much of the damage usually comes from the way the storm pushes up the sea to very high levels ahead of the wind so that there is widespread water damage to whatever is in the path of this storm surge, particularly if the storm comes ashore at high tide. Orson added about 3 metres to the height of the sea at the time it crossed the coast; this would have been a disaster if high tide (normally about 5 metres) had been taking place. However, the normal tide at that time was only about 1.6 metres, so the coastline could cope with the extra water.

The Darwin hurricane

This storm was not a special event, but is a typical image of weather disturbances in that area of Australia at that time of year. The well-known tropical cyclone Tracey (25 December 1974) which devastated Darwin was not tracked on radar, since the system was not in operation at that time.

Worksheet: Weather storms

Four storms are available in RAPIC images for investigation. These are:

- the Melbourne storm
- the Sydney storm
- the tropical cyclone Orson (near Dampier)
- the Darwin storm

Pictures are recorded every 10 minutes. The screen width for Melbourne and Sydney storm pictures shows 256 km, and for Darwin and Orson shows 512 km.

RAPIC is the name of the system used by the Australian Bureau of Meteorology to observe rain and hail. The coloured images show the intensity of the radar reflections. In general the more rain the more reflections, so the colours tend to show the heaviness of the rain.

Colour	Type of rain	Amount of rain
Black	nil	0 to 0.2 mm per hour
Dark blue	light showers	0.2 to 2 mm per hour
Light blue	medium rain	2 to 10 mm per hour
Yellow	heavy rain	10 to 20 mm per hour
Green	very heavy	20 to 40 mm per hour
Pink	thunderstorm	40 to 100 mm per hour
Red	very heavy thunderstorm and hail	over 100 mm per hour

For each storm you study, use the following questions are a guide.

1 How many video stills are used for each storm? Convert this to real time.

2 How far did the storm travel in the time displayed? How fast did the storm move?

3 In which direction did the storm appear to move for most of the time?

4 Amount of rain. Use the area size to approximate the area for each colour and hence calculate approximately how much water was dumped by each storm (in litres or megalitres).

5 Use these figures to write a newspaper article so that the average reader can understand the figures. For example, how many Olympic-size swimming pools of water is this?

Additional questions on cyclone Orson

Orson was one of the most powerful cyclones ever to affect the coast of Western Australia. A cyclone is a large system that rotates around a central 'eye' which is quite calm. Can you see the eye on the video?

We consider the eye as the centre of the storm, as the winds around it are very fast indeed. The pressure at the centre of the 'eye' of Orson was the lowest ever recorded in an Australian cyclone. *The reason the tape stops is that the storm blew away the radar tower.*

6 In which direction (around the eye) are the winds rotating?

7 How wide is the area of calm, i.e. within the 'eye'?

8 In which direction did the cyclone appear to move for most of the time?

9 The large-scale map on the right shows the path of the cyclone. The dots show its position at 9 a.m. each day. The lines of latitude are 5° apart. This corresponds to a distance of about 555 km.

10 Estimate the distance travelled by Orson on each day (17 to 24 April) and hence its average speed.

11 You can see a clear track in a westerly direction and then a curve towards the south. Estimate the direction (as a bearing) of the cyclone's path on each day.

12 As it moved towards the coast it got faster. (You should see the cyclone speeding up. Is this evident on the video?) The insert map shows the path of the cyclone from midnight on the 22nd to 10 a.m. on the 23rd. It crossed the coast at 4.30 a.m. Calculate its speed over water and over land.

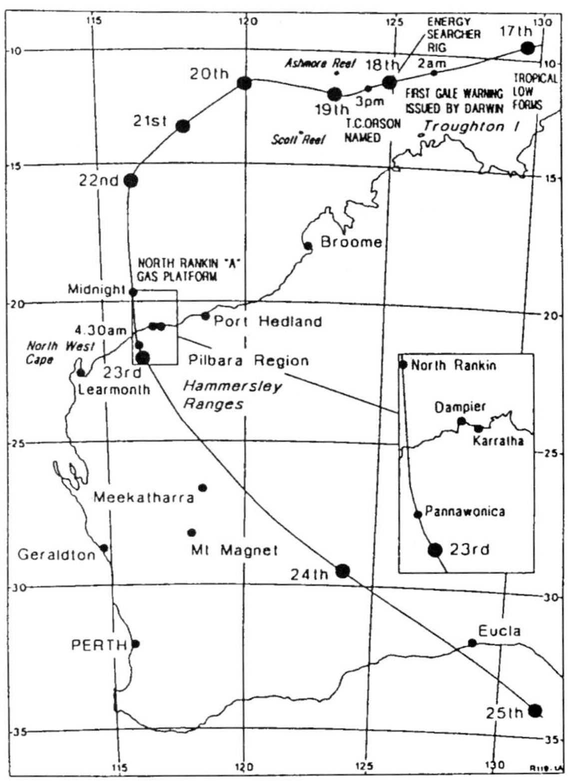

5 Fitness data

All my students had heard that triathletes are among the fittest of athletes, but few had any idea of how their fitness might be measured. So watching Sandra, Mark, Suong and Huynh wired up on the exercise bikes and collecting data while they rode was a very instructive, as well as enjoyable, way to study statistics.

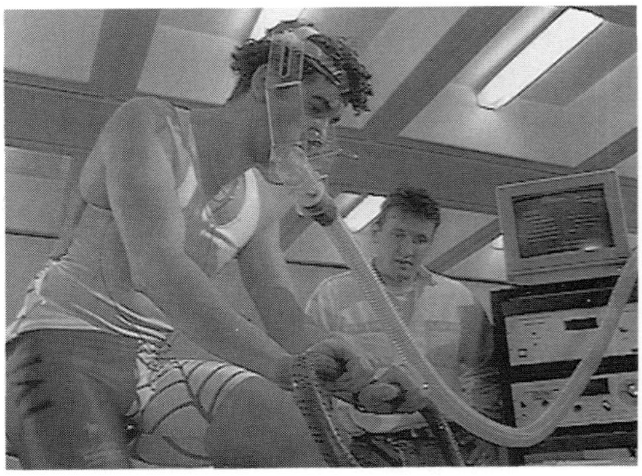

The two statistics used here are heart rate (HR — in beats per minute) and volume of oxygen uptake (VO_2 — in litres per minute).

Each person rode for 3 minutes (at a fixed speed) and then was measured for another 2 minutes during recovery. The results can be read directly from the screen but are also presented in the following results table (in 15-second intervals).

	Huynh		Suong		Mark		Sandra	
Exercise time	VO_2	HR	VO_2	HR	VO_2	HR	VO_2	HR
0.00	0.50	80	0.48	82	0.52	60	0.51	63
0.15	0.78	85	0.79	90	0.65	80	0.75	81
0.30	0.95	92	0.98	98	0.78	85	0.79	90
0.45	1.05	125	1.05	129	0.88	110	0.82	108
1.00	1.25	135	1.28	138	0.92	119	0.91	120
1.15	1.32	150	1.35	140	0.99	122	0.95	128
1.30	1.35	160	1.39	152	1.11	132	1.11	132
1.45	1.38	161	1.40	162	1.15	135	1.18	135
2.00	1.36	162	1.38	163	1.20	140	1.23	142
2.15	1.34	161	1.40	165	1.20	140	1.22	142
2.30	1.35	160	1.41	164	1.20	140	1.23	143
2.45	1.37	163	1.40	165	1.19	142	1.21	143
3.00	1.35	162	1.42	166	1.21	141	1.23	142
Recovery time								
0.15	1.14	158	1.25	160	1.18	135	1.20	139
0.30	1.00	150	1.18	150	1.11	130	1.15	130
0.45	0.98	146	1.01	143	1.05	122	1.10	119
1.00	0.85	140	0.94	139	1.00	110	0.99	107
1.15	0.80	135	0.83	137	0.98	99	0.95	96
1.30	0.78	129	0.79	129	0.92	89	0.90	86
1.45	0.77	120	0.75	122	0.82	79	0.83	73
2.00	0.75	110	0.72	119	0.79	71	0.79	69

The data told us many 'stories' as we analysed and graphed it. For example:

- Are there differences between the trained athletes (Sandra and Mark) and the students (Suong and Huynh)? Graph and compare the recovery rates for each person.
- Are there different patterns in the oxygen uptake data compared to heart rate? Why is a high oxygen uptake ability a good thing?
- Which had the greatest rate of increase when exercise started, heart rate or oxygen uptake? All my pupils predicted it would be heart rate.

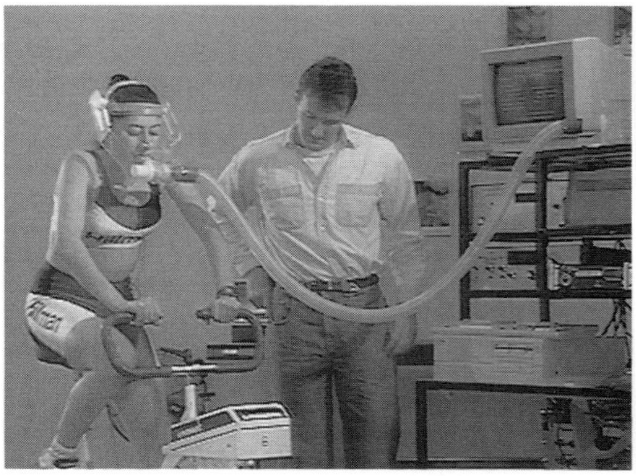

6 Tennis

Wimbledon, England

One wag at Wimbledon always knew that tennis was boring! — and argued it by calculating that the ball is only in play for just 9.5 minutes in a two and a half hour game. A spokesperson for women's tennis seized on this calculation and claimed it proved women's tennis gave more value for money.

One of my students brought in this whimsical newspaper article and we saw the opportunity to use video data to find out if the claims had any mathematical truth. So we obtained some footage from the final games of the men's and women's 1991 Australian Open tennis finals.

1 I gave out a copy of the article and showed a snippet of part of one game. From the snippet pupils *estimated* the percentage or fraction of time the ball was in actual play.
2 Then we figured out how to collect the data we needed to solve this question. The electronic clock on the screen and the pause button were invaluable aids, although a stopwatch could also be used.
3 We calculated the results over several games for both the men's and women's finals, and then extrapolated these for a five-set match.
4 My class also made an estimate and then calculated from a sample the total distance a player would run or walk in the duration of a five-set match. They were quite surprised by how far it was.

It proved a great exercise in data collection and organisation and generated very spirited discussion.

It was a great deal of fun, but also showed us clearly how much raw data there is available in the videos of many sports events.

Population centre of Australia

Years 8 to 12

If Australia had only 20 people, ten in Perth and ten in Brisbane, the 'population centre' would be halfway between them. This rather fanciful idea is the basis of the concept of the population centre. The stimulus of an American newspaper article leads to interesting data analysis, to historical and social trends, and to the useful mathematical tools of weighted averages and modelling.

Features

Teaching and learning

- Estimation
- Historical population trends
- Social issues
- Link to geography
- Problem-solving strategies
- Computer use
- Group work

Major mathematical content

- Weighted averages
- Latitude and longitude
- Distances (Pythagoras or tabled data)
- The mathematical modelling process

Mathematics content is further explored in the notes in stages 6 and 7.

Equipment and preparation

- Copies of the five worksheets as necessary — one per pupil (Worksheet 2 has figures up to 1989 and more recent data may well be available from the ABS.)
- Blackboard ruler and some suitable materials to develop the concept of weighted averages
- Test the computer program if available

Expected timing

- From three to seven sessions on the investigation

A teacher's story

I saw a somewhat offbeat article (see Worksheet 1) about the American population centre and thought it could be a stimulus to explore the statistics of population census. This led to one main investigation and three interesting extensions. We covered many maths ideas which linked up with other subjects, knowledge of our country, problem solving, group work and the power of computers.

I found the article very useful in triggering the students' imaginations. It is strange how sometimes the offbeat point of view can be a stepping stone to worthwhile learning. After reading the article, the challenge of estimating the current 'centre' of Australia on the map tapped their ideas. Comparing the different perceptions in the class was illuminating and a trigger to find the 'right answer'. I also had them estimate the populations and the latitudes and longitudes of some of the largest cities in Australia, 'testing' their geographical knowledge.

I followed this up by providing the data they had just estimated (Worksheet 2), including historical data. This led to interesting discussion about population trends in Australia.

I felt it necessary to spend some time on the mathematics of weighted averages (Worksheet 4). This provided ideas on how to solve the problem and clarified the meaning of the 'centre'. Eventually I let small groups wrestle with the main problem using whatever ideas they had. There is always a tension between 'teaching' the algorithm, problem solving and letting students cope with the concept.

Finally we had a good discussion about the modelling assumptions that were being made when we used only a few cities to estimate the populations. (Even the 'answer' provided by the Australian Bureau of Statistics has assumptions built into the model.)

Comments

I like the problem because it can be done at increasing levels of sophistication. All the groups got an answer, and since no one knows the 'correct' answer, there was much interest in the various methods used.

Using the main problem as a starting point, several independent extensions (Worksheet 5) were possible:

- the same problem on a state or territory basis — Where is the current centre within Queensland?
- a historical perspective — Is the current centre 'moving'? Where has it been in the past? How fast is it moving?
- a social issues perspective — Did the American article allow for the indigenous population? What does this mean in the Australian context?

Summary of the stages in this investigation

1 **Introducing the concept of population centre**

2 **Estimating the current centre**

3 **Estimating populations and positions of the cities**

4 **Providing and discussing the data**

5 **Group problem solving**

6 **Problem-solving strategy 1: related problem**

7 **Problem-solving strategy 2: simplify the problem**

8 **Group presentations**

9 **Examining the modelling assumptions**

10 **Extensions**

1 Introducing the concept of population centre

I handed out Worksheet 1 with the article. I linked this to their memories of the 1991 census.

The computer people clearly had fun with the United States statistics. Read the article and then think about the similar idea for Australia.

Sometimes a little clarification is needed about the concept. For example, if there was just one person in Perth and one person in Brisbane, where would the centre be? (Obviously halfway between.) What if there were ten people in Brisbane and just one in Perth? (Much closer to Brisbane.)

Somebody, pondering the person mentioned in the article, made the comment attributed to Andy Warhol about 'everyone being famous for 15 minutes'.

2 Estimating the current centre

We had a brief discussion of the American data, with its steady shift to the west, and the reasons for this. Then I wanted to shift the focus to the Australian context.

For the whole of Australia, place a cross on the map on Worksheet 1 where you think the current Australian population centre is.

Compare your estimate with that of a neighbour. Do you want to change your estimate?

Students were very keen to see each other's guesses. The guess represents a student's initial perception and generates interest and motivation to see how accurate it is. The small discussions that erupted were a great way for students to clarify their thinking, to share and gain knowledge of their own country, and to begin explaining how they would solve the problem.

3 Estimating populations and positions of the cities

Before handing out the data needed to solve the problem I asked them to estimate the populations and positions of the largest cities in Australia.

Place dots in pencil on Worksheet 1 to show what you think are the 12 largest cities in Australia. Next to each put your estimate, in thousands or millions, of the population of each of those cities.

Also estimate the latitude and longitude of each of the 12 cities, and write them on the map next to each city in pencil.

The difficulty of this task can be adjusted by supplying some of the information (e.g. the names of the cities). Although quite hard, it turned out to be a very worthwhile activity. I allowed plenty of discussion, and changing of dots and numbers.

I found that students needed a brief reminder of the meanings of latitude and longitude:- that latitude increases downwards and longitude increases to the right.

4 Providing and discussing the data

I handed out Worksheet 2 which contains a map and the 1989 census data for populations of the 12 largest cities and some others of interest. The students then corrected their estimates.

The worksheet also contains the data for several periods in the past. What can you learn about population trends in Australia?

I allowed several minutes for the data to be read and discussed informally. Students found the data fascinating. In particular they could see when Canberra did not exist, and that there has been a move into the big cities from smaller country towns.

5 Group problem solving

Find the current Australian population centre, as accurately as possible.

The problem is very open-ended. I decided to let groups tackle it by themselves, if they wanted to, and offered the others some hints (see stages 6 and 7).

They worked on this for some time. I challenged students to present their report in a style similar to the newspaper article. Since the groups finished at different times, the three extension problems were available (Worksheet 5) for them to go on with.

The computer was used by some students who had trouble with the arithmetic. (Only 70% of the total population is included in the computer data, so some assumptions must be made about the rest.)

I like the open-endedness of the problem, but there are lots of uncertainties that arise and groups may need a bit of a helping hand.

A large state map (showing southern New South Wales) is probably sufficient to pinpoint the answer to the nearest town, but local district maps are available from map shops if more accuracy is required.

A reasonable assumption is that the rest of the population is spread around the country towns evenly. However, there are some areas where very few people live.

The answers

The Australian Bureau of Statistics now has our urban centres broken up into geographical regions of 100 m squares. They know the location (accurate latitude and longitude) of each square and how many people live in the square. If you assume everyone lives in the centre of the square, you could put all this data through a computer program. Stimulated by the American article, the Bureau of Statistics did just that for this investigation. The answer (based on 1989 data) was latitude 34°04' South and longitude 144°36' East. The nearest town is probably Booligal (NSW) (33°52'S, 144°53'E).

Date	Latitude	Longitude	Nearest town (all in NSW)
1989	34°04'	144°36'	Booligal
1986	34°05'	144°44'	Booligal
1971	34°21'	145°03'	Hay
1954	34°36'	145°30'	Carrathool
1911	34°55'	145°32'	Gunbar

Although not as dramatic as the US data, these show a definite drift north and west. Since 1989, there has been a large population shift towards Queensland, so the suspicion is that currently the centre would be moving north-east. The 1991 census figures might confirm this.

The accuracy of the earlier results is a little suspect due to the uncertainty of past census procedures.

The extensions encourage pupils to link the calculations to an awareness of current social trends and the reasons behind them. Worksheet 2 gives the populations of the largest cities for several different years. The percentage that these represent of the total Australian population in the same year is also given. (The Aboriginal population was not included in the official census until 1951. For more accuracy, some assumptions must be made about this. The information in the *Appendix* could be copied for students.)

6 Problem-solving strategy 1: related problem

It is unlikely that your students will need all the hints and assistance that the following two sections offer. Try to save for them something that makes this a problem-solving challenge.

I felt that many of my students would not have any idea of how to tackle this problem because they had no great experience with the idea of weighted averages. So I offered them Worksheet 4 as a set of hints.

Find the average of 10, 10, 10, 2.

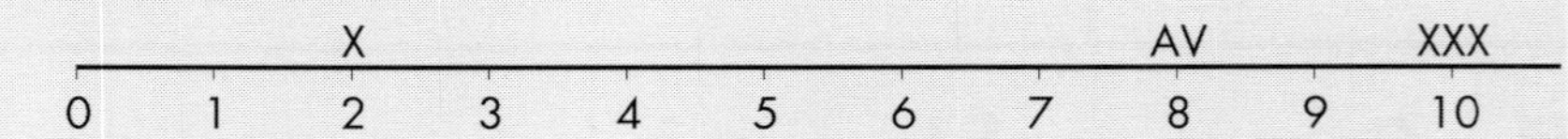

Using an algorithm, the 'balance point' is: $\frac{10 \times 3 + 1 \times 2}{4} = \frac{32}{4} = 8$

After working on the problems they all had diagrams showing weighted averages on a number line.

How do these diagrams relate to the populations and finding the centre?

There are at least two methods available. One splits the distance between the cities in the inverse ratio of the populations. On the number line above, the position of the average divides the distance between 2 and 10 in the ratio 3 to 1. If students use this method, they will need hints about how to include more than two cities as the calculations are quite complex.

The simpler method involves using latitude and longitude directly. Most students had some idea of the relationship, but I felt that they needed more specific ideas on how to use the latitude and longitude in the calculation. However, your students may feel that this is enough of a hint.

7 Problem-solving strategy 2: simplify the problem

The following is a graded set of hints in the form of simplified problems that gradually become more complex and develop the algorithm. I prepared them all, but did not need to use them. You could try to stop giving hints at the point where your groups have some ideas about how to proceed but leave enough problem solving to give them a feeling that they are actually solving the problem for themselves, and not just number crunching.

To give you some ideas about how to solve the problems using all 12 cities, we will engage in a well-known strategy to help solve a difficult problem: *simplify the problem*.

For a very simple problem to develop the algorithm, try problem 2 from Worksheet 4. This has just ten people in Newcastle and one person in Perth.

I chose the problem of two cities with the same longitude (roughly): Hobart 43°S (200 000) and Townsville 19°S (100 000).

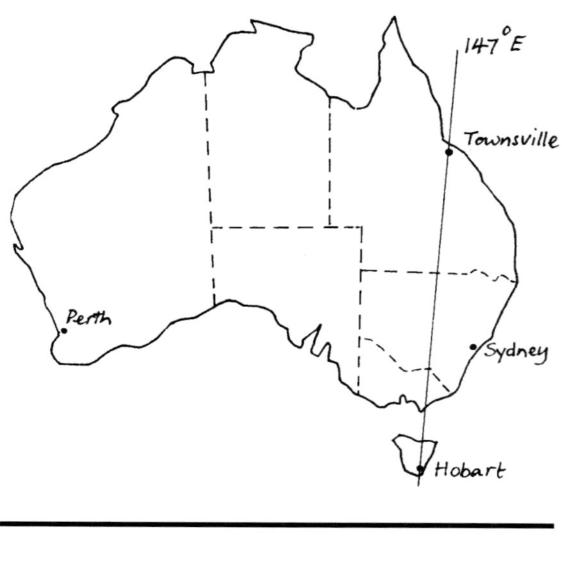

Two cities

Method a

Find Hobart and Townsville on the map. They have about the same longitude (147°). How many degrees of latitude are between them? (Answer 24°)

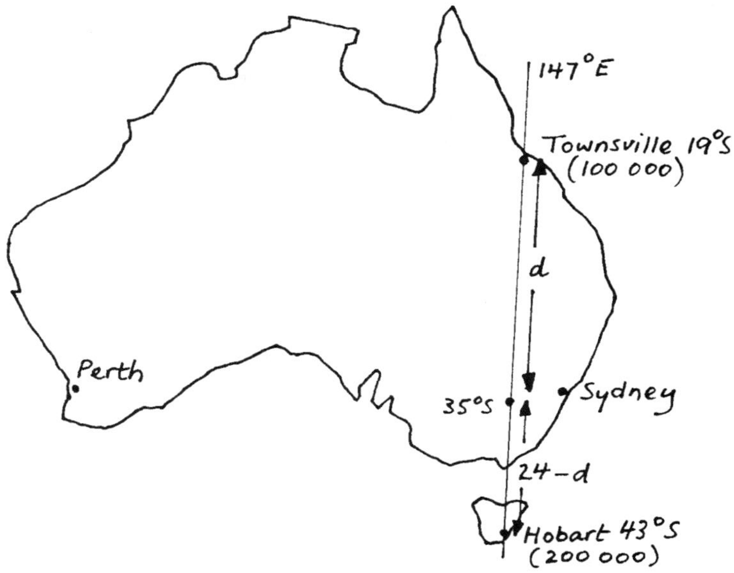

The centre position is:

$$100\,000d = 200\,000(24 - d)$$
$$300\,000d = 4\,800\,000$$
$$d = 16$$

The centre should split the 24° in the ratio 2 to 1, and should be closer to Hobart; that is, 8° from Hobart and 16° from Townsville, at 35° S.

Method b

The direct weighted average is:

Latitude:

$$\frac{200\,000 \times 43 + 100\,000 \times 19}{300\,000} = \frac{10\,500\,000}{300\,000} = 35$$

To make this clearer, use a blackboard ruler and weights such as text-books. Balanced two weights at one end and one at the other end. The ruler could be balanced by holding it at one point about a third from the heavier end.

I found that a see-saw analogy was useful to students. We imagined two people (weighing 100 kg each) at one end of a plank, and one at the other end weighing 100 kg. The plank would balance one-third of the way from the heavier end. The plank would enable the three people to be replaced by a weight of 300 kg at the balance point.

An interesting aside (from the world of physics and friction) is that if you place one finger under each end of the ruler, and slide them inwards, your fingers will finish up at the balance point. Try it with a softball bat.

Adding a third city

We now add in a third city to Hobart and Townsville, say Darwin 70 000 at 12° South and 131° East. The problem is still a two cities problem since Townsville and Hobart have now been 'replaced' by a single point of 300 000 balanced at 35° South and 147° East.

We need to find the weighted average of both the latitude and longitude.

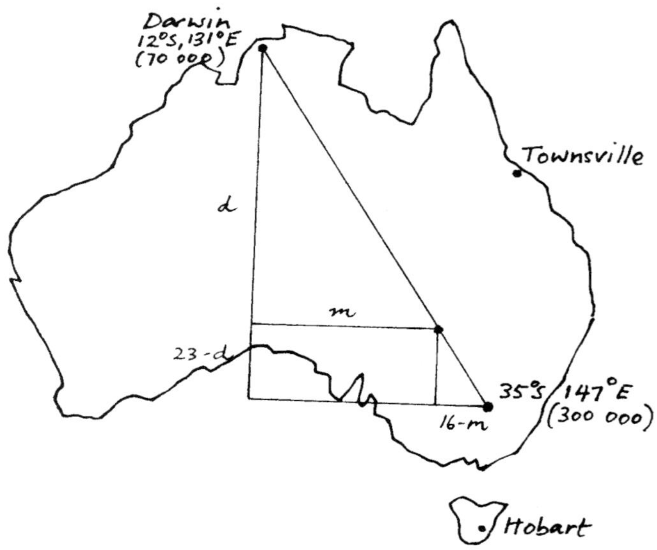

Method a

Latitude:

$$70\,000 \times d = (23 - d) \times 300\,000$$
$$70\,000d = 6\,900\,000 - 300\,000d$$
$$370\,000d = 6\,900\,000$$
$$d = 18.6$$

Longitude:

$$70\,000 \times m = (16 - m) \times 300\,000$$
$$70\,000m = 4\,800\,000 - 300\,000m$$
$$370\,000m = 4\,800\,000$$
$$m = 13.0$$

Hence the new 'centre' has population 370 000 at 30.6° South, 144° East.

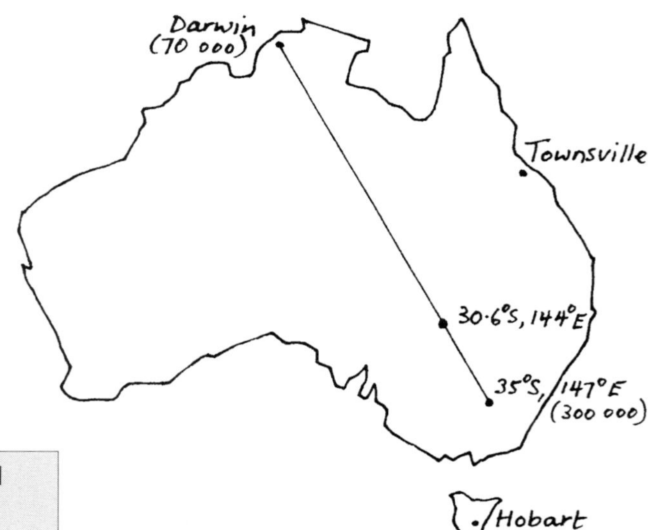

Method b

We could weight all three latitudes and longitudes (using rounded figures and hundred thousands).

Longitudes:

$$\frac{147 \times 2 + 147 \times 1 + 131 \times 0.7}{3.7} = 144.0° \text{ E}$$

Latitudes:

$$\frac{43 \times 2 + 19 \times 1 + 12 \times 0.7}{3.7} = 30.6° \text{ S}$$

> Clearly this algorithm is quick and efficient, but I did not want to reach this point without pupils understanding the concept.

Computer program demonstration

There are several computer versions of this problem. I had the Excel version on a Macintosh, and the program *Centre of city populations*. Here is how it works.

1 Open both the SS and map files.

2 Type 'h' next to Hobart, and 't' next to Townsville. All other spaces in this column should be blank. This will tell the program to display only these two cities on the map, and to show the position of the population centre of those two for the year shown (change this to 1989 if necessary).

3 It will be seen that the population centre is about a third of the way up from Hobart, and the exact latitude (and longitude) is given at the top of the spreadsheet.

4 Now extra cities (for example Darwin) can be added one by one, noting the effect of each.

8 Group presentations

Each group was asked to present their report in turn.

A large wall map was used after the centre had been calculated on the worksheet. Students used the large map to close in on the actual centre. Then an atlas showing south-west NSW in detail was used to find the nearest town to the predicted centre.

9 Examining the modelling assumptions

It was important for students to know that the model only gives an impression of the 'true' answer; they need to realise the approximate nature of modelling.

Whenever we model we make assumptions, the essence of mathematical modelling. For example, you have been given cities that have only about 70% of the total population. Where is the other 30% and how can these be considered?

The discussion rambled over many areas so I made a summary on the board. The following points were included. It was important to let students hypothesise about how they would cater for these assumptions.

What are the assumptions we have made that have allowed us to use this weighted averages model?

My pupils wanted to know what effect our recent school excursion to Alice Springs would have on the 'centre'.

1 The most important issue is the approximate nature of the data. The census counted all the people who were located in each centre at the time of the count but things have since changed. Even if we could know exactly where each person's house was, there could still be some argument about the part of the house in which they actually lived! No matter how accurate the data, the model must be approximate.

2 The total population is incompletely covered as it is not possible to include all the people in a small set of data. We have chosen to limit the data to the places where 70% of the people live. Are the rest spread out more or less evenly? In 1911, the ABS in their calculations had to use 101 urban centres (all those over 3000) and this still only covered 55.1% of the population. Today, the eight largest cover over 70%. For the ABS calculation in 1986, they used the centroids of all 1338 statistical local areas across Australia.

3 Over time, the boundaries of cities have changed. Cities may actually have more people if more urban sprawl is included. In 1851 what are now suburbs of Sydney and Melbourne were quite separate towns.

4 We can critically revisit the original article from a social perspective. For example, in 1790 was the American native population considered and if so how? Early census data are perhaps not as accurate as today's and certain assumptions have been made.

5 The early census data were taken at different times.

The early data for Australia include estimates of the Aboriginal population based on major regions. In the early nineteenth century about 80% of the Australian population was Aboriginal. These have been left out of the official count. (In some cases they actually counted them to know how many to leave out!)

For more detail, see the *Appendix* or the book *The Aboriginal Population of Australia* by L.R. Smith (ANU Press 1980). You could copy this for interested students.

10 Extensions

Three extensions are provided on Worksheet 5.

- Consider the same problem on a state or territory basis. For example 'Where is the current centre within Queensland?' (Worksheet 3 is needed for this.)
- Look at the historical perspective. Is the current centre 'moving'? Where has it been in the past? How fast is it moving? Recognition of the mobility of the population and insight into the reasons for this show why census data are needed to plan and provide services for the shifting population.

> The American centre has moved many hundreds of kilometres over time. The Australian centre has been much more stable. My class figured the centre is currently moving about 8 metres per hour; that is, the length of our classroom in 1 hour.

- Investigate a social issues perspective. Did the American article allow for the indigenous population? What does this mean in the Australian context? In particular the absence of Aborigines in the data until the 1950s staggered the students, and led to an exploration of the decline and subsequent rise of the Aboriginal population in Australia. Use the *Appendix*.

Appendix: Aboriginal population and the census

> This information is summarised from *The Aboriginal Population of Australia* (L.R. Smith, ANU Press 1980).

Before the time of white settlement, the Aborigines comprised the total population of Australia. For a variety of reasons, most were not counted as full members of the Australian society until the census in 1966. To quote Smith:

> If we consider 'the census' to be the final printed report, then it can be said that the determination of 'who is an aborigine' for 'census' purposes has become increasingly complicated and increasingly error-prone over the years. (p. 13)

Census taking was the responsibility of the states before 1911. As a result there developed a diversity of approaches about who was included in the count. For example, sometimes Aborigines employed by whites on stations were counted, in other cases only those with more than half 'white blood' were included.

Based on traditional information from Aborigines themselves, and reports from white people at the time, Smith has estimated that size of the Aboriginal population in 1788, and in the years before it, was about 314 500. The table below shows how it is estimated that the population was spread across the different state areas.

From 1861 to 1891 the total numbers of Aborigines has been estimated. This is the best estimate that can be made of the numbers of Aborigines who are *not included* in the official census counts (Smith, p. 209). After 1900 the estimates are of the number of 'full-bloods'; these people were definitely not included in the census counts until 1966.

Date	NSW	Vic.	Qld	SA	WA	Tas.	NT	Aust.
1788	48 000	1 500	120 000	15 000	62 000	4 500	50 000	314 500
1861	16 000	2 400	60 000	9 000	44 500	18	48 500	180 400
1871	12 000	1 700	50 000	7 500	40 000	85	44 000	155 285
1881	10 000	1 200	40 000	6 346	35 500	120	38 500	131 666
1891	8 280	900	32 000	5 600	31 000	139	33 000	110 919
1901	3 778	271	25 137	4 385	29 000	0	23 048	85 668
1911	2 022	196	20 000	4 000	24 023	3	21 756	70 000
1921	1 597	144	14 014	1 609	25 587	0	17 439	60 300
1933	1 299	992	12 532	1 988	25 623	0	18 643	60 101
1947	953	208	9 100	2 139	20 338	0	13 900	46 638
1954	1 405	141	13 824	2 501	10 195	0	15 505	43 572
1961	1 490	253	13 656	2 147	10 121	0	17 386	45 053

Worksheet 1

USA's population center: Missouri farm

by Carol J. Castaneda
USA TODAY

Source: Census Bureau

By Julie Stacey. USA TODAY

Smack dab in the middle of Garland Eaton Sr.'s farm, just up from the Crooked Creek and by the old Keysville Lookout Tower, is the new population center of the USA.

The spot, determined by a USA TODAY computer analysis of 1990 Census figures, is where the country would be balanced if each of the nation's 248.7 million residents weighed the same.

As of April 1, 1990, the day the Census was taken, the spot was in Crawford County, Mo., between Keysville, with fewer than 20 residents, and Cherryville, a small town where old-timers sip coffee at the general store-post office.

These days, the talk is turning to Eaton and his centrally located 818-acre plot.

'Everybody knows Garland, it's going to be the talk of the town,' says Cherryville Postmaster Cary Schwelch.

The center — 80 miles southwest of St. Louis — is farther south and west than ever.

'It's moving southwest because the population has been going to Florida and California,' says Larry Taylor, a geographic specialist with the Census Bureau. 'A lot of people move toward the sun.'

That trend has been going on since the first Census in 1790, when the center was 23 miles east of Baltimore.

The Census Bureau, which uses the same formula used by USA TODAY to come up with the population center, hasn't released its 1990 location because it's still trying to decide whether to adjust the Census.

If the Census is adjusted, the center will move again, a little farther south and west.

'I wouldn't think it would move more than a mile or two,' Taylor says.

And what does Eaton think of being at the center of it all?

'I'm not famous,' he says. 'I'm just a country boy.'

Worksheet 2

Populations of largest cities

Cities	1989	1986	1971	1954*	1911*	1851*
Sydney (33.8°S, 151.1°E)	3 623 600	3 472 700	2 807 828	1 863 161	698 800	53 294
Melbourne (37.8°S, 145.0°E)	3 039 100	2 931 900	2 503 450	1 524 111	628 400	23 143
Brisbane (27.5°S, 153.1°E)	1 272 400	1 196 000	867 784	502 320	145 000	—
Adelaide (34.9°S, 138.4°E)	1 036 700	1 003 800	842 693	483 508	196 600	14 577
Perth (32.0°S, 115.9°E)	1 158 400	1 050 400	703 199	348 647	116 000	2 513
Newcastle (33.0°S, 151.8°E)	419 100	416 100	351 536	187 144	12 816	—
Canberra (35.3°S, 149.2°E)	276 900	257 850	158 880	28 277	—	—
Wollongong (34.4°S, 150.9°E)	235 900	232 500	199 048	90 852	4 725	—
Hobart (42.9°S, 147.3°E)	181 200	179 000	153 216	95 206	27 505	23 107
Geelong (38.0°S, 144.3°E)	145 200	125 833	122 087	72 595	21 630	8 291
Gold Coast (28.1°S, 153.4°E)	198 300	184 270	66 697	—	—	—
Townsville (19.3°S, 146.8°E)	100 500	106 300	71 265	40 471	13 678	—
Launceston (41.5°S, 147.2°E)	92 420	90950	62 241	49 303	20 937	10 855
Ballarat (37.6°S, 143.9°E)	80 210	78 260	58 620	48 030	38 686	1 500
Toowoomba (27.6°S, 152.0°E)	81 071	78 140	59 524	43 149	16 160	—
Broken Hill (32.0°S, 141.5°E)	22 580	25 170	29 808	31 351	30 953	—
Bendigo (36.8°S, 144.3°E)	68 050	64 950	45 936	36 918	17 883	800
Darwin (12.4°S, 130.9°E)	72 900	56 482	31 899	8 071	944	—
Total Australia	16 833 085	16 018 400	12 937 200	9 090 395	4 455 005	406 657
Percentage included in cities above	72	72	71	60	45	33

* Aborigines not included.

Worksheet 3

Populations of largest cities in each state and territory 1986

NSW (70%)	**Lat °S**	**Long °E**	**5 531 526**
Sydney	33.83	151.07	3 472 700
Newcastle	32.95	151.78	255 787
Wollongong	34.32	150.88	206 803
Central Coast	33.42	151.35	162 669
Maitland	32.73	151.57	43 247
Wagga Wagga	35.12	147.38	37 577
Albury	36.05	146.95	37 164
Tamworth	31.08	150.93	30 729
Orange	33.27	149.12	28 935
Dubbo	32.25	148.63	25 796
Lismore	28.82	153.28	24 896
Broken Hill	31.95	141.47	24 460

Vic. (70%)	**Lat °S**	**Long °E**	**4 160 856**
Melbourne	37.82	144.97	2 931 900
Geelong	38.02	144.33	125 833
Ballarat	37.55	143.87	63 802
Bendigo	36.75	144.28	53 944
Shepparton	36.35	145.43	30 238
Melton	37.68	144.60	24 394
Warrnambool	38.35	142.50	22 706
Wodonga	35.05	146.95	20 918

Qld (65%)	**Lat °S**	**Long °E**	**2 624 595**
Brisbane	27.47	153.10	1 196 000
Gold Coast	28.07	153.42	163 332
Townsville	19.28	146.80	96 230
Toowoomba	27.57	151.97	71 362
Cairns	16.90	145.75	54 862
Rockhampton	23.37	150.50	54 362
Mackay	21.15	149.18	38 603
Bundaberg	24.88	152.35	33 368
Mount Isa	20.73	139.48	23 348
Gladstone	23.85	151.27	22 033
Maroochydore	26.67	153.12	20 635
Maryborough	25.55	152.70	20 177
Caloundra	26.78	153.13	16 215
Hervey Bay	25.22	152.83	14 410
Tewantin-Noosa	26.38	152.12	11 296
Gympie	26.22	152.72	10 772
Beenleigh	27.72	153.22	10 344
Nambour	26.60	152.98	9 579

SA (70%)	**Lat °S**	**Long °E**	**1 382 550**
Adelaide	34.92	138.36	1 003 800
Whyalla	33.03	137.58	26 900
Mount Gambier	37.82	140.78	20 813
Port Augusta	32.56	137.80	15 291
Port Pirie	33.17	138.03	13 960
Murray Bridge	35.10	139.30	11 893
Port Lincoln	34.73	135.87	11 552

WA (70%)	**Lat °S**	**Long °E**	**1 459 019**
Perth	31.95	115.85	1 050 400
Rockingham	32.32	115.75	30 635
Bunbury	33.33	115.65	23 031
Kalgoorlie	30.72	121.47	22 232
Geraldton	28.77	114.60	21 726
Mandurah	32.52	115.72	18 016
Albany	35.00	117.88	16 316
Port Hedland	20.30	118.58	13 069

NT (70%)	**Lat °S**	**Long °E**	**154 421**
Darwin	12.40	130.87	56 482
Alice Springs	23.68	133.87	22 759
Katherine	14.45	132.25	5 691
Nhulunbuy (Gove)	12.18	136.77	3 515
Tennant Creek	19.63	134.70	3 503

Tas. (65%)	**Lat °S**	**Long °E**	**446 473**
Hobart	42.88	147.32	179 000
Launceston	41.45	147.15	66 286
Devonport	41.17	146.37	22 645
Burnie	41.05	145.92	20 665
Kingston	42.98	147.33	10 932
Ulverstone	41.15	146.18	10 055
Bridgewater	42.73	147.20	8 654
New Norfolk	42.78	147.07	6 152
George Town	41.08	146.85	5 315
Wynyard	40.98	145.75	4 705
Queenstown	42.05	145.96	3 593
Smithton	40.83	145.15	3 414

ACT (96%)	**Lat °S**	**Long °E**	**258 910**
Canberra	35.28	149.15	257 850
Hall	35.17	149.07	289

Worksheet 4

Although the answers are provided below, your challenge is to verify these.

1 For each of these sets of numbers:

- Estimate the average.
- Work out the average with a calculator.
- Sketch a number line that shows the numbers in the set and the average.

a 10, 10, 10, 2 **b** 10, 10, 4 **c** 20, 20, 2 **d** 2, 2, 2, 2, 2, 20
e 30, 30, 3 **f** 30, 3, 3 **g** 30, 10, 10, 10 **h** 4, 4, 4, 20, 20

2 Where is the centre of population for these situations? (Note that all data are rounded to keep the numbers simple.)

a ten people in Newcastle (32°S, 132°E), one in Perth (32°S, 116°E)
b ten in Hobart (43°S, 147°E), five in Townsville (19°S, 147°E)
c 20 in Hobart, ten in Perth (locate this centre on a map)

3 Find the centre of the populations for these cities.

a Same latitude: Sydney (150°E) 4 000 000, Perth (115°E) 1 000 000
b Same latitude: Sydney (150°E) 4 000 000, Adelaide (139°E) 1 000 000, Perth (115°E) 1 000 000
Method: Combine the answer to **a** with Adelaide.
c Same population, different latitudes and longitudes. Make a sketch.
Brisbane (27°S, 153°E) 1 000 000, Perth (32°S, 116°E) 1 000 000
Method: You need the average of both latitude and longitude.
d Different population. Make a sketch.
Melbourne (39°S, 145°E) 3 000 000, Adelaide (35°S, 139°E) 1 000 000
Method: You need the weighted averages of both latitude and longitude.

4 For three cities, make a sketch and place your answers on a map.

a Brisbane (27°S, 153°E) 1 000 000, Sydney (34°S, 150°E) 4 000 000, Perth (32°S, 116°E) 1 000 000
b Perth (32°S, 116°E) 1 000 000, Darwin (12°S, 131°E) 70 000, Townsville (19°S, 147°E) 100 000
c Choose any three cities from Worksheets 2 or 3.

Method a

Firstly find the population centre (X) of Brisbane and Sydney (which have the same longitude).

X's population is 5 000 000 and its latitude is $\dfrac{1 \times 27 + 4 \times 34}{5} = ?$

Then find the population centre of this new centre (X) and Perth.

Its population is 6 000 000 and its longitude is $\dfrac{1 \times 116 + 5 \times 150}{6} = ?$

Its latitude is about 32.5°S.

Method b

Weight all the longitudes and latitudes.

Longitudes: $\dfrac{1 \times 116 + 1 \times 150 + 4 \times 150}{6} = ?$

Latitudes: $\dfrac{1 \times 32 + 1 \times 27 + 4 \times 34}{6} = ?$

5 Find the population centre for these four cities:

Townsville (19°S, 147°E) 100 000
Hobart (43°S, 147°S) 180 000
Darwin (12°S, 131°E) 70 000
Perth (32°S, 116°E) 1 200 000

Answers

1 a 8 **b** 8 **c** 14 **d** 5 **e** 21 **f** 12 **g** 15 **h** 10.4
2 a (32°S, 149°E) **b** (35°S, 147°E)
c (39.3°S, 136.7°E)
3 a 5 000 000 at 143°E
b 6 000 000 at about 142°E
c 2 000 000 at 29.5°S, 134.5°E
d 4 000 000 at 38°S, 143.5°E
4 a 6 000 000 at 32.5°S, 144.8°E
b 1 170 000 at 29.7°S, 119.5°E
5 1 550 000 at 31.5°S, 122.3°E

Worksheet 5

1 Find the current centre for each state and territory. Use Worksheet 3.

a First mark a personal estimate on your map for each state or territory of interest.

b Ask some others to do the same and discuss any differences in perceptions.

c Use the data on Worksheet 3 to locate the current centre.

d Write a short report including any surprises, and your predictions about movement trends.

2 a Look again at Worksheet 1. How has the centre moved over time for the USA as a whole? In what direction is the centre moving and why?

b Use an atlas to estimate distances and calculate how fast the American centre 'has been moving'. Comment on this, given California's current population of 45 million people. When did this migration start?

Date	Latitude	Longitude	Nearest town (all in NSW)
1989	34° 04'	144° 36'	Booligal
1986	34° 05'	144° 44'	Booligal
1971	34° 21'	145° 03'	Hay
1954	34° 36'	145° 30'	Carrathool
1911	34° 55'	145° 32'	Gunbar

c Use the table above to calculate the speed of movement for Australia over different time periods. Work it out in metres per hour.

d What is the direction of movement of the Australian centre? What does this mean?

e Why is it important to recognise these shifts in population? What groups in the community need to know such information?

3 a Did the census take into account indigenous peoples? How would this affect the calculations? Use the *Appendix* notes (obtain from your teacher).

b Census procedures in earlier times were often defined in different ways. Is it likely the true centre of population in America in 1790 was centred in New Jersey? (California was not settled until well into the 19th Century — the 'West' was not at all well known to those in the East.)

c Read the *Appendix* notes. Use the table of data and make some assumptions about distribution to calculate the Australian population centre in 1788. (The First Fleet had 777 convicts and about 400 other personnel.)

d Are there any Australians who might not be recorded in our 1991 census?

Top of the class

Years 9 to 12

Is averaging a set of scores to find the best performer always successful? Even if it works, is it fair? What are the other possibilities? Do they always work and are they fair? The use of statistics to rate people and its validity is under trial in this investigation.

Features

Teaching and learning

- Deals with a realistic problem
- Computer use to test properties of number sets and create new ones
- Group discussion and debate
- Reporting

Major mathematical content

- Properties of the mean, median and sum of ranks
- Linear scaling of scores, by graph and formula
- Standard deviation, resulting in *z*-scores
- Recognition that all statistics have strengths and weaknesses, and that the most suitable for a given situation is a matter of debate and value judgment.

Mathematical content is further discussed in the *Content epilogue*.

Preparation and equipment

- One copy of Worksheet 1 for each group of five students, cut it into strips
- Students will find calculators and possibly a computer useful

The software program *Top of the class,* if available, has been written to make the analysis in this investigation as easy as possible.

Expected timing

- Three or four sessions

A teacher's story

Statistics play a huge part in the rating of human performance — in sport, in business, in education. This is especially pertinent for students as ratings largely determine their future careers. The most glaring example of this is in tertiary selection. There is always healthy debate on the particular statistics that are used and their fairness. I felt it was important not only to learn the techniques of particular treatments such as averages, ranking and *z*-scores, but also to be able to debate the fairness of each.

How do you find the best performer when you have a number of scores to combine? Most students will say to find the average, which is the method most schools use to find the 'top of the class'. However, as the investigation demonstrates, this is often most unjust.

In order to help students grasp the concepts involved, I deliberately constructed some simple sets of scores with interesting properties. I used a computer program to do this, and a further extension invites students to invent other such sets using the same program. These sets of scores were used by students to debate ideas of fairness and the suitability of the proposed method for finding the 'top of the class'.

Students may become disenchanted with the ability of any statistic to meet a particular need. This should lead to a healthy scepticism, and a distrust of statistics not carefully justified. Questions that should be asked at all times include: 'Who chose this statistic as the best way to make this decision? Is it fair, or does it favour the party who chose it?'

Comments

The students greatly enjoyed being on the other side of the fence and playing the role of the judge rather than the 'victim'. It certainly gave them a healthy scepticism about the validity of statistics and a realisation of the need to carefully choose and justify why a particular statistic is used. It stimulated their awareness, and for the following weeks they often found examples in the paper or on the news or sports programs of a great variety of uses of statistics in rating performance.

Summary of the stages in this investigation

1 Introducing the problem

2 Group debate: who is the best?

3 Report on findings and methods used

4 Alternative 1: sum of rank orders

5 Does 'sum of ranks' always work?

6 Inequities in ranks based on averages

7 Alternative 2: linear scaling

8 Does linear scaling always work?

9 Dealing with outliers

10 Alternative 3: *z*-scores

11 Do *z*-scores always work?

12 Consolidation

13 Extensions

1 Introducing the problem

Because the Olympics had been on TV a lot recently, I asked about the methods for scoring the performers in various sports. Two good examples are the diving and the long jump. The winner of the long jump is the athlete who makes the longest single jump out of six tries. Yet this is not so in the diving. The winner has the highest average score on ten dives, but may not necessarily perform the best single dive. Why is this system used? How could it have evolved? Is it fair?

I explained that sport is an example of a more general problem: to find the best method of combining several scores for different tasks in order to find the best performer.

Students named a lot of sports and how the scoring systems work to determine the winners. We certainly built up a picture of the immense variety of the way statistics are invented and applied.

I ran the MCTP activity on the *Mathematics of diving* which was an excellent prelude to this investigation.

What are some other situations when a set of scores have to be combined to find the best or to rate people?

This question is designed to discover whether they clearly have the general concept of the problem. I wanted this awareness, so that they realised that the particular context we were about to explore would be seen as representative of a wider set of problems.

I made no apology for using test scores as the basis of the investigation (rather than say, sports). After all, senior students are so much the 'victims' of statistical analysis of their performance and the effect it has on their future choices. I felt they should be aware of some of the related issues.

2 Group debate: who is the best?

This part worked in groups of five. I used Worksheet 1 and cut set 1 scores into five strips, giving one to each group member.

These are the scores that five different people got for three tests in mathematics. Imagine you have to recommend tertiary entrance on the basis of these results. They have to be combined so that a single number will tell us the order of ability of the five people.

In your group, discuss and debate, and decide who you think is best, second best and so on. You will have about 15 minutes for this task, then be ready to report both on your answer and reasons. Will each group come up with the same ratings?

In this way each pupil 'owned' a set of scores. In the discussion that followed they were keen to 'prove' that their scores were the best. This 'ownership' really invigorated the debate.

A most lively debate followed. They took it very seriously. They knew the numbers were contrived for a purpose, but the potential importance of such a task was accepted by all. We ensured that each group had a reporter who was taking some notes.

Most groups immediately calculated the average — but as these were all the same they were forced to consider other possibilities. I tried to stimulate them to think laterally without suggesting any particular method.

3 Report on findings and methods used

Firstly I asked each group to write their five best performances in rank order on the board.

Can group 3 tell us how they decided Erin was the best?

There was intense interest as each group's results went up and immediate murmurings about the differences.

Secondly, when all group ratings were up, I invited the class to ask any group to justify their choices. As each group reported I tried to find common ground and look for the good mathematical points, which I wrote on the board.

One group decided that the median was another type of average, and they would try that. They had difficulty justifying the choice of the middle of three scores as the single measure of performance overall, but their real trouble came when they found that the medians of the scores in set 1 were all the same!

4 Alternative 1: sum of rank orders

You have seen the varied way statistics can be used and know how important the outcomes can be. I would like to show you three alternatives that mathematicians have invented and might be useful. These are:
- **sum of rank orders**
- **linear scaling**
- ***z*-scores**

I found their willingness to pay attention to these statistical methods much greater because of the earlier debate and their role of having to make 'judgments'.

I gave out the set 2 results from Worksheet 1, again in strips, one for each member of the group.

Let's try this new idea: put the scores for each person in each subject into rank order (best gets 1, lowest gets 5), and add the ranks.

Work out the ranks for tasks *B* and *C*, and then total the rankings for each person.

On task A the ranks are:

Angelo	1
Barbara	2.5
Chad	5
Davina	2.5
Erin	4

Note that Barbara and Davina shared 2nd and 3rd ranking in test A by both scoring 8. So the rankings of 2 and 3 are 'averaged' between them at 2.5.

Chad's sum of ranks is 8.

The others are 9, 8.5, 9 and 10.5.

So Chad is the winner!

But he came bottom in test A!

5 Does 'sum of ranks' always work?

Try this method on set 3 (give out strips) and on set 1. Also work out common averages (or means).

We found that, in set 3, the averages were all different, but the sum of rankings were all equal (9 each). In set 1, both the averages and the sum of rankings are the same.

6 Inequities in ranks based on averages

I wrote this set of scores on the board for their consideration.

Do averages work here?

Yes, the order is *a* to *e*.

Which task contributes most to the average? Does the score in task B contribute anything to the final order? Does task C contribute as much as task A?

Person	Task A	Task B	Task C
a	10	7	5
b	8	7	5
c	6	7	6
d	4	7	7
e	2	7	7

Clearly task B could be any score (all scoring 1 or even 10 for all students) and it still would make no contribution to the final order. The idea we are after is that the *range* of scores is important. So clearly task A unduly influences the final order. Is this fair? Surely all task results should have an influence on the final order?

What would happen if we organised the scores so that tasks A and C had the same spread?

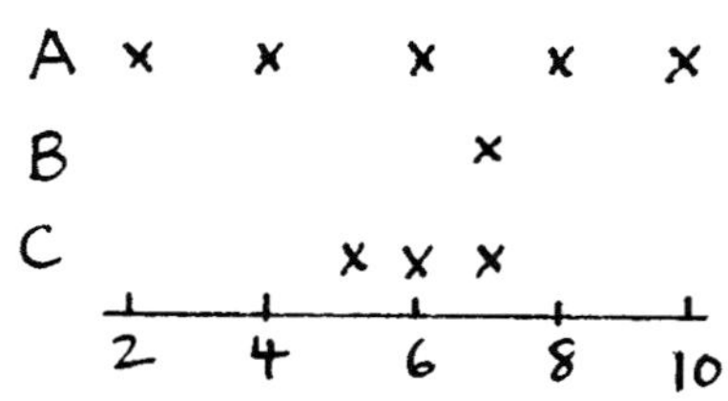

I explained the idea as follows: The simplest method here is to change the scores for task C. We changed the lowest score to the same as the lowest score in A (that is 2) and the same for the highest (10). The middle score is halfway (in this case) and so will become 6. Below is what we now had on the board.

Person	Task A	Task B	Task C	Unscaled average	Unscaled rank	Scaled C	Scaled average	Scaled rank
a	10	7	5	7.3	1	2	6.3	3
b	8	7	5	6.7	2	2	5.7	5
c	6	7	6	6.3	3	6	6.3	3
d	4	7	7	6	4	10	7	1
e	2	7	7	5.3	5	10	6.3	3

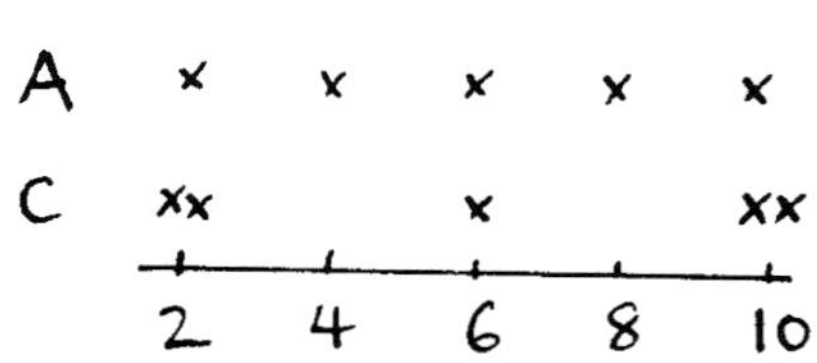

The order has changed dramatically, since task C now makes as much contribution to the final order as task A. Task B still makes no difference.

7 Alternative 2: linear scaling

It was time to teach a more formal approach to linear scaling, since they now understood why it was 'fairer' than using unscaled scores, and had the basic idea.

I demonstrated the graphical method.

You start with a square on graph paper with side-length 10, the total marks possible. You draw a straight line from bottom to top to meet the bottom at the lowest score obtained and the top at the highest score. Then you can read the graph to convert actual scores to scaled scores.

We tried it for the scores on the board. Here are the results.

Person	A	Scaled A	C	Scaled C
a	10	10	5	0
b	8	7.5	5	0
c	6	5	6	5
d	4	2.5	7	10
e	2	0	7	10

The usual approach is to make the range of each set of scores the greatest possible, and to spread the in-between scores proportionally or linearly.

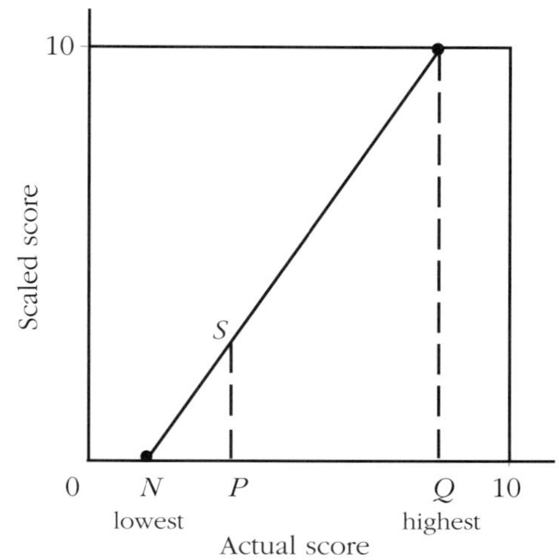

The scaled score on the vertical axis is the height of the triangle in this diagram, *PS*. This is a fraction of the total height of 10, and is the same as the fraction that *PN* (the base of the small triangle) is of *QN* (the base of the larger triangle).

So the formula is $PS = 10 \times \dfrac{PN}{QN}$

or $\quad \text{scaled score} = \text{total marks} \times \dfrac{\text{score} - \text{lowest}}{\text{highest} - \text{lowest}}$

It was quite clear why the method didn't work for the task B scores; the line was vertical!

Some students were able to follow the derivation of the formula.

8 Does linear scaling always work?

I asked student to apply this method to set 3.

Explore each of the methods we have tried:

- **mean**
- **sum of ranks**
- **average of scaled scores**

on this new set.

Then try the average of scaled scores method for sets 1 and 2. Be ready to report your results.

This may be a stage where some classes could stop and possibly explore some of the extensions. However, my class seemed interested and able so I pressed on.

For set 1, scaling makes all students equal! For set 2, scaling produces a sensible order. For set 3, means produce an order, ranks are equal, but scaling the scores produces almost the opposite order. The conclusion we are looking for is that the choice of method requires a value judgment. In real life this choice may be made by someone who is using statistics to their own advantage. We need what is called 'healthy scepticism'.

9 Dealing with outliers

Task	Student *a*	*b*	*c*	*d*	*e*
A	10	9	8	7	0
B	2	4	5	6	7

Take a look at these scores for two tasks. One student failed to hand in any work for task A. How does this affect the outcome? Is it fair?

Here are the answers we obtained together.

	Student *a*	*b*	*c*	*d*	*e*
Average	6	6.5	6.5	6.5	3.5
Scaled A	10	9	8	7	0
Scaled B	0	4	6	8	10
Scaled average	5	6.5	7	7.5	5

If student *e* had obtained any score for task A, the spread of scores for task A would be much smaller. The linear scaling process means that task A would then have more influence on the final order than it can using these scores.

Discussion about fairness focused on the undue influence of the 0 on the scaling process.

What should we do to be fair?

We 'gave' *e* a score of 5 on test A, and saw the difference it made.

The basic idea to be raised here is that outliers, such as the 0, have too great an influence on the linear scaling process. They could be ignored, but then you have to make difficult decisions about what constitutes an outlier and what does not. The best solution is to use a measure of spread that does not give the outliers undue prominence.

I used this occasion to introduce the most commonly used measure of spread — *standard deviation* (sd).

On your calculator you have a key to find 'σ_{n-1}'. This is called the standard deviation, and is a way of calculating the spread using all the results, and not depending only on the extreme values.

I showed them how to use their calculators to find standard deviation values and their means. We practised on the sets above.

A	10	9	8	7	0	sd = 3.96	mean = 6.8
B	2	4	5	6	7	sd = 1.92	mean = 4.8

I sketched the following graph to make the meaning a little clearer. It shows how the standard deviation measures the spread each side of the mean.

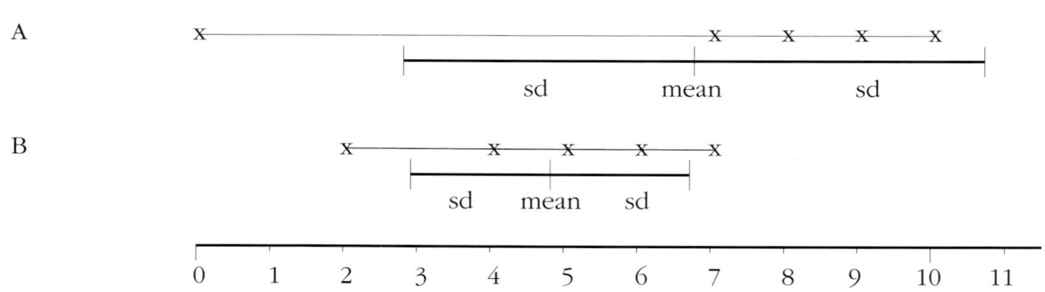

10 Alternative 3: z-scores

At this stage I went the whole distance, and showed them how to use this new measure of spread to standardise the actual scores into *z*-scores.

The standard deviation can be used as a unit to measure the spread of each score each side of the mean. The way this is done is with a simple division, comparing the difference between the score and the mean with the standard deviation. The result is called the *z*-score.

$$z\text{-score} = \frac{\text{score} - \text{mean}}{\text{standard deviation}}$$

The example made it clearer.

A *z*(A)	10 0.8	9 0.55	8 0.3	7 0.05	0 −1.7	sd = 4	mean = 6.8
B *z*(B)	2 −1.4	4 −0.4	5 0.1	6 0.6	7 1.1	sd = 2	mean = 4.8
Sum of *z*	−0.6	0.15	0.4	0.65	−0.6		

The graph helped also. It showed how the *z*-scores had almost made the spreads equal.

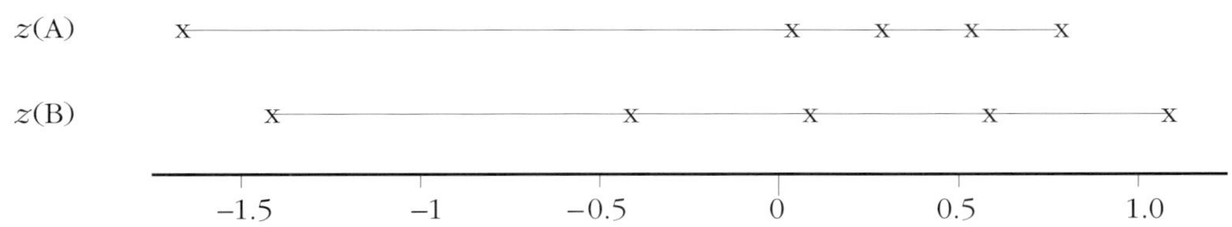

11 Do z-scores always work?

At this stage I set the groups the task of analysing the score sets 1, 2 and 3 using this new method.

As they probably expected by now, although *z*-scores make sense for some sets, for set 1 they produce equal rankings.

When the calculators gave standard deviations of 0 for the set of equal scores I had to remind them what it meant: a measure of the amount of spread. Clearly dividing by 0 is meaningless, so *z*-scores are not possible for these results. It is sensible to make them all 0, since the equal scores will make no contribution to the sum of *z*-scores.

I concluded the main part of the investigation with a summary of the major idea of this investigation: many different statistics may be chosen to help to make decisions. However, which one you choose to use may depend on who you want to win! It is always wise to ask why a particular choice has been made, and to find out who is benefiting by the choice.

12 Consolidation

Worksheet 2 has two sets of results for ten students, with five and eight subjects, as well as an indication of some tasks that could be undertaken for practice in the skills learned. It is expected that students will use calculators or the computer program *Top of the class* to help with this task.

Although they seem more typical of real results, they each have at least one special property. What is it?

13 Extensions

A: Sport scoring

I suggested that students develop a project on the topic of scoring for diving, skating and other sports. The basic idea is that several judges give a score and these different scores are combined. How do they do it? Is there any method of avoiding the pitfalls we have encountered in our own investigation?

> The method uses averages, but there is often a rule that the two extreme scores must be dropped first. Students will probably understand why after stages 6 and 9 above.

B: Invent your own sets of numbers

The special sets of numbers in Worksheet 2 were carefully chosen to have special properties. They are not, of course, the sort of thing that usually happens. However, it is a valuable challenge to find similar sets of numbers with the same properties. Students could choose some of the following challenges. The computer program *Top of the class* can help with this.

Find some results (out of 10) for three tasks for five students which have different numbers, with:

1 Equal medians
2 Equal means
3 Equal sums of ranks
4 One order for means, which is reversed for the averages of linearly scaled scores
5 Equal standard deviations
6 Equal z-scores

C: Standardisation procedures for TES in your State

Each State has grappled with the task of combining Year 12 scores to form an all important tertiary entrance score (TES) in its own way. In most cases some standardisation is used. For senior students, learning about this is not only valuable statistical training, but has personal significance. I made it a project, but for those who took it on I provided documentation from the body responsible for doing this work.

Content epilogue: background information for teachers

Means, and the impact of spread

The method of combining scores by adding, or finding an average, is so common that it is wise to challenge it, as we have done in this investigation. Suppose you grade one task from 2 to 10 and give most kids either 7 or 8 for another task, of equal importance educationally. If you just add the scores, will it give a sensible total that reflects the equal importance of the two tasks? Of course not, since the first task will have a much greater influence on the total. The spreads should be roughly equal.

In sports this direct method is modified to prevent any one judge having an unfair influence on the score by giving very high or very low scores to any one competitor; they simply have a rule that both extremes are ignored before the average is calculated. Is this fair? It is probably as good as we can get, because in the sport the final score is expected soon after each competitor has performed. Methods that require knowledge of the ranking among the scores for each judge, and either the total range or the standard deviation of all the scores from one judge, in order to find the competitor's score, are clearly impossible.

The idea that the spreads should be roughly equal is one of the main ideas in this investigation. A technique used by generations of teachers has been the idea of converting the sets of marks to new scores that have the same spread using linear scaling. This method is useful and both a simple graphical method and a simple formula can be used for the conversion. However, outliers have undue effect on the result.

Another familiar method over the years has been the use of rank order only, since the rank orders will have the same mean and overall spread. This is a very simple method, and is probably to be preferred. This method does not reflect the differences between the actual scores, but these may not be significant. The outliers are given no special prominence in this method.

The other major idea in this investigation is that statistics are designed for a purpose, and each method has its strengths and weaknesses. Some of them may fail to be useful or cannot even be used under certain special circumstances, even though they may be very helpful in most cases. A caveat has been stressed; beware the use of statistics designed to produce a message of benefit to the powerful. The well-known book *How to lie with statistics* (by D. Huff) is very supportive of this general message.

Standard deviation and *z*-scores

The use of standard deviation to measure spread is well-known. The standard deviation is the square root of the variance, which is obtained by adding the squares of the differences between the actual scores and their mean, and dividing by *one less than* the number of points involved. The standard deviation is thus a measure of the total amount that the scores differ from their mean. The diagram may help to clarify this idea.

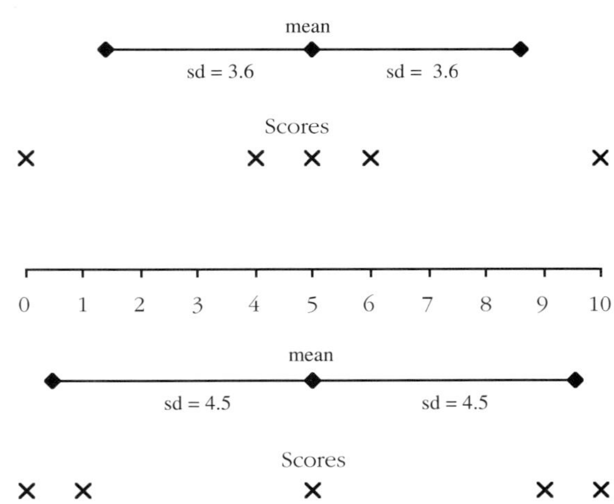

The formula reflects this aim:

$$\text{standard deviation} = \sqrt{\frac{\Sigma\,(x - \bar{x})^2}{n - 1}}$$

where x is the score, $\bar{x}$ the mean, and n the number of scores that form the mean.

Once we have a measure of the spread, the actual scores can be allocated *z*-scores. These just show the number of standard deviation 'units' that each score is above or below the mean. So all you do to find the *z*-score is find the distance from the mean, and divide by the standard deviation.

$$z\text{-score} = \frac{x - \bar{x}}{\text{sd}}$$

Clearly the *z*-score can be positive for scores over the mean, and negative for the others, and their total should be zero! Although *z*-scores have many uses, some associated with tables of normal distributions, these difficult ideas are not needed to understand the *z*-scores themselves.

Worksheet 1

Copy once per group of five, and cut this into strips.

Set 1: Angelo	A 10	B 2	C 6
Set 1: Barbara	A 9	B 3	C 6
Set 1: Chad	A 8	B 4	C 6
Set 1: Davina	A 7	B 5	C 6
Set 1: Erin	A 6	B 6	C 6
Set 2: Angelo	A 10	B 2	C 6
Set 2: Barbara	A 8	B 3	C 7
Set 2: Chad	A 3	B 8	C 9
Set 2: Davina	A 8	B 8	C 2
Set 2: Erin	A 7	B 8	C 2
Set 3: Angelo	A 10	B 3	C 6
Set 3: Barbara	A 8	B 5	C 5
Set 3: Chad	A 5	B 7	C 4
Set 3: Davina	A 3	B 4	C 10
Set 3: Erin	A 0	B 6	C 9

Worksheet 2

Here are some scores for students in a number of tasks. Determine their order. 'Sum of' has been represented by Σ.

Student	Task A	Task B	Task C	Task D	Task E	Mean	Σranks	Σscaled	*z*-scores
a	5	50	56	74	45				
b	32	47	50	84	58				
c	95	38	42	67	50				
d	66	50	41	56	44				
e	45	59	50	76	11				
f	86	70	43	50	35				
g	50	54	44	62	32				
h	50	54	44	62	32				
i	29	19	56	62	50				
j	67	50	47	49	91				

Student	Task A	Task B	Task C	Task D	Task E	Task F	Task G	Task H	Mean	Σranks	Σscaled	*z*-scores
a	100	30	47	72	40	75	30	47				
b	90	38	43	60	20	65	48	70				
c	61	36	40	45	41	55	62	80				
d	63	32	51	90	30	70	47	35				
e	56	55	41	82	45	40	49	41				
f	80	45	49	64	65	45	38	20				
g	23	47	45	55	60	80	32	60				
h	40	35	52	70	56	20	60	65				
i	85	40	60	40	28	51	55	30				
j	72	54	50	10	25	35	66	75				

Dying for a smoke

Years 9 to 12

This investigation uses real data from the Australian Bureau of Statistics and the Anti-Cancer Council of Victoria to explore the relationship between smoking and deaths by lung cancer, chronic bronchitis and heart disease. It is an attempt to let the data speak for themselves; no 'preaching' is needed. Students' attitudes are developed over time and a wealth of information about likely consequences of certain actions is not, in itself, going to change habits long established.

Features

Teaching and learning

- Interpretation of survey results about a socially sensitive area
- Demonstrates how statistical models can be created to produce information and to make predictions about the future
- Enhances student awareness of the impact of smoking on death rates

Major mathematical content

- Analysis of the outcomes of a survey, using means and distributions
- Search data for relationships between variables
- Interpretation of data in terms of rates and relationships to age and gender
- Calculation of relative risks and attributable risks (that is, probabilities) for defined populations
- Interpretation of the results in terms of statements about social behaviour of individuals
- Creation of a sample and questionnaire for a survey

Mathematical content is further discussed in the *Content epilogue*.

Preparation and equipment

- Copies of the worksheets
- The book *Cancer Issues Resource: An Activity Book for Post-Primary Teachers* (see *Content epilogue*) is a considerable help with discussing the issues of cancer with secondary students
- Someone outside of the school, such as an experienced doctor or nurse may be able to help
- Access to a good database that also draws graphs or a statistical program for the analysis of the survey results (extension) is a great advantage

Expected timing

- Between two and ten lessons depending on the depth of treatment

A teacher's story

In these investigations I tried to link the realities of community use of the drugs, tobacco and alcohol, to the dangers they present, particularly leading to deaths. There are several aims:

- for students to use mathematics and computers to learn more about their world;
- to challenge the easy-going assumption that there are no negative consequences of smoking — students realised that many early deaths of adults in their community were due to smoking;
- to show students how statistical data can provide information for predictions about the future. The number of deaths each week due to smoking is a given fraction of the total number of lung cancer deaths. This fraction, the 'attributable risk', can be worked out mathematically;
- to show how students can use data available in the community to learn about significant trends that affect us all. The investigation called on data from several sources to produce conclusions that speak loudly of the dangers involved.

I knew that this could be a sensitive area. It is important to know whether any students have lost close friends or relatives recently due to any of these causes; if so, tread carefully. I found that sensitive handling of the issue can help the whole class share this event, and so provide understanding and support. Such a circumstance can create a climate in which very real life and death issues can be discussed openly, and mathematics can be seen to make a valuable contribution to the discussions.

Students have been exposed to much advertising from the tobacco and alcohol industries and there is a clear need for some solid facts and sensible discussion. There is no support for smoking in this investigation; the facts do not allow it. The students were very interested, and for once they were sure that the mathematics lesson was going to be relevant to their lives. Background information is given in the *Content epilogue* to help you to tackle this sensitive issue.

Comments

I had to work quite hard to overcome some resistance to dealing with such issues. 'It's an old people's problem.' One avenue was to emphasise the increasing role mathematics and mathematicians are playing in what is a very serious social debate. I used articles we gathered from newspapers and magazines to emphasise this point. Who collected the data? Who used it to make projections? Ultimately, it was their commitment to and interest in the mathematics that took us over the initial hurdles and finally led them to talk seriously about the topic.

The advertising debate was very active at the time we did this work, and I used that as a stimulus to consider the effects of smoking. Should packs be marked with even more explicit warnings against smoking? Should the tobacco industry be allowed to sponsor sporting events?

Summary of the stages in the investigation

1 **Smoking and lung cancer: an introduction**

2 **Who smokes?**

3 **Lung cancer trends**

4 **Lung cancer: gender and age differences**

5 **Chronic bronchitis and heart disease**

6 **Extensions**

1 Smoking and lung cancer: an introduction

I held a short discussion which aimed to expose students' present attitudes and awareness of tobacco as a cause of death. I also asked all students to write a short paragraph to state their own attitudes towards this matter.

This sequence of lessons does not aim to preach against the supposed evils of smoking. Instead, it aims to let students explore the related statistics and draw their own conclusions.

Students were aware of the government anti-smoking initiatives (for example the QUIT campaign). However, the facts (estimates of deaths) are more hard hitting. We talked about the effects of advertising on our behaviour and about how the causes of deaths are not generally discussed.

We discussed a general plan for how we might investigate the matter further. The first step involved studying the statistics relating smoking and lung cancer deaths. We considered the possibility of a school survey of smoking habits and attitudes. This is described later as an extension activity.

I was staggered to find that many students simply did not believe that smoking could be a cause of lung cancer.

The debate about the effects of passive smoking, and the ethical questions about allowing people the freedom to choose to smoke or not, were a basis for discussion.

The government had just raised the tax on tobacco, yet again, and was talking about requiring extensive warnings on cigarette packs. Students wanted to know why it was such a big deal.

2 Who smokes?

I handed out Worksheet 1, showing the results of a survey of smoking habits of teenagers. This large national survey was conducted in 1990, and run by the Anti-Cancer Council of Victoria. Worksheet 1 has the major tables and graphs from an article by the authors of that survey.

I explained the procedure used for the survey.

A *stratified two-stage sample* of 12- to 17-year-olds was taken in this way:

- Firstly, 351 secondary, and some primary, schools were randomly selected from all states and 88% agreed to participate.
- Secondly, 20 students from each school were randomly selected from each of Year 7 to Year 12. Over 26 000 students were involved in total.

I organised the students into groups to work out a verbal summary of the data on Worksheet 1. These reports were to include observations of differences between age groups and between the sexes, and were to indicate trends.

After about 10 minutes, I asked several groups to contribute their reports about what the data were saying. I also asked them to tell me whether they thought smoking was now more common or less common than in 1990.

3 Lung cancer trends

I handed out Worksheet 2, which shows in graphical form the number of deaths due to lung cancer and also the total tobacco imports to Australia, from 1910 (the earliest available complete data) to 1990.

The dramatic rise in lung cancer deaths is very clear; there is also a dramatic rise in tobacco imports. This does not, in itself, provide evidence of cause and effect, as the tobacco industry is very quick to point out. (There have been other very dramatic rises, such as population and car ownership, over the same period; without further evidence any of these *could* be the cause of the changes in lung cancer deaths.)

Again I used the data for interpretation and discussion of the implications.

What story does the graph tell you? Can you relate the graph shape to other things that were happening at the time?

When did lung cancer start to become a problem?

The line graph on Worksheet 2 shows how annual tobacco imports have changed between 1910 and 1990.

Is there any possible relationship between cigarette consumption and lung cancer deaths? Does this *prove* that cigarette smoking causes lung cancer deaths?

I challenged my pupils to find and bring in samples of the very many newspaper articles relating to this topic. We discussed these with particular emphasis on the way mathematics played a part in the article.

4 Lung cancer: gender and age differences

I handed out Worksheet 3 and the students answered questions 1 to 5. About 30% of men in Australia are active smokers, as are about 27% of women. The cause of death of each person is listed on their death certificate. Lung cancer is the most common cause of cancer death for men and the third most common for women.

There is also a graph on the Worksheet 2 relating to data from the United States. (Comparable data are not available in Australia.) This shows how the rate of males dying from lung cancer (in the age group 55 to 64) is affected by two things: the age at which a smoker starts, and the number of cigarettes smoked per day.

Which age groups suffer most from death by lung cancer? Why do men get lung cancer more than women?

Taking men who start smoking before age 15, what is the rate at which they die from lung cancer if they smoke:

- **from 10 to 20 cigarettes per day?**
- **from 21 to 39 cigarettes per day?**

How do these rates compare with the rate of men who do not smoke?

What does this graph tell you about the way teenage smoking affects your chances of dying from lung cancer?

Why would the government be concerned about the effects of cigarette advertising on young people?

5 Chronic bronchitis and heart disease

This section deals with two other types of death where smoking is implicated.

Chronic bronchitis is a very significant cause of death in Australia, and is closely tied to cigarette smoking. It has nothing to do with cancer. A graph of age-related death rates and other information are found on Worksheet 4. Cigarette smokers are more than five times as likely to die of chronic bronchitis than are non-smokers.

Heart disease is the *leading cause of death* in Australia and smoking is one of its major causes. Worksheet 4 contains some background information on types of heart disease.

Worksheet 4 is deliberately more open than the others. Students are challenged to analyse the information on the worksheet and ask themselves what 'story' is conveyed by the data.

6 Extensions

A: School survey of smoking

1 Plan a survey: the sample

I pointed out the steps involved in running a survey: identifying the sample to be asked, writing, testing and modifying the questions, conducting the survey, displaying and finally interpreting the results to answer the original question. I introduced them to the process called 'stratified random sampling'.

To deduce information about different groups in the population we need to select at random some members of each group. Which groups will we try to include?

How can we randomly select these people?

2 Plan the questionnaire

Groups of students worked on preparing their proposals about the questions for presentation to the class. When different proposals were ready, each group copied their ideas for the whole class, and explained their ideas publicly.

Finally a set of questions was agreed on by the class as a whole.

Despite the important message about the dangers of smoking, the Anti-Cancer Council is finding that people's smoking behaviour is not changing as fast as they expected. Why is this?

What questions can we ask that might help us work out why smokers don't quit even when they know the consequences of keeping on smoking?

3 Test and modify the questions

We did a trial run with a small number of people who would not be included in the final sample. If possible, the same test should be given again to the same people after a break of 1 or 2 weeks. If the test is reliable, the people will make the same responses both times.

After examining the responses, and possibly talking to the people about what they thought the questions meant, some questions may need to be revised.

4 Conduct the survey

For the actual running of the test we planned methods and standard procedures that would increase the chance of honest answers. (For example, for the national survey described above, teachers were absent, and the anonymous answer sheets were placed in plain envelopes.)

Before the actual survey took place, I did some clearing of the ground first, by getting permission of the principal and staff of the school. I also found it useful to get the cooperation of the school council. Then the class was involved in the administration of the survey according to the agreed procedures.

5 Tabulate and graph results

There is a vast amount of data involved in a survey such as this. Keeping track of all the responses and tabulating them needs careful organisation, which I required of the students.

A database program or statistical package is the best way to treat the data when it is finally in table form. Each person asked is a 'record', and the responses to each question are placed in each 'field'. The database program can search out the records with any specific responses to any questions, and also find those that belong to two or more fields at once (e.g. 13-year-olds who smoke, but less than one cigarette per week).

A good database program will also be able to graph the data in several ways. For example:

- the frequency of the values in any field (a stem-and-leaf plot, box-plot or histogram);
- comparisons of the distributions of the values on two fields, also including mean values;
- a scatter-plot to show the relationships between values in two fields.

6 Interpret the results, and compare with national results

I split the task of interpreting the data among groups, giving different aspects to each. (To improve later discussion, I arranged for each section of the interpretation to be tackled independently by two groups.)

Once the students have some conclusions, they can compare them with the large-scale national survey described earlier.

You must stress the need to be cautious here. Students have a tendency to jump to conclusions that are not supported by the data. Every statement of conclusion must be carefully reasoned. The frustration of not having asked the right questions may occur.

B: Comparing smokers with non-smokers: relative risks

I handed out Worksheet 5, and we carefully went over the mathematics involved.

The *relative risk* is a mathematical measure that uses data about causes of death and relates them to the habits of the person who died. For example, for lung cancer, the relative risk of smoking compares the rates of death by lung cancer for smokers and non-smokers.

Answers to Worksheet 5

	Relative risk Males	Relative risk Females
Ever smoked	6.9	6.8
Heavy smokers	17.45	16.5
Current smokers	9.1	8.6
Ex-smokers	4.4	4.2

Of the people who die of lung cancer, the rate among those who *have ever been* smokers is nearly seven times the rate among those who have never smoked. That is, the ratio of the death rates from lung cancer for smokers to non-smokers is 6.9. This is called the 'relative risk of smoking for lung cancer'.

The relative risk depends on the amount of tobacco the smoker consumes. Considering just the people who are smokers when they die, the relative risk is almost ten; that is, there is a ten-fold increase in lung cancer risk caused by smoking up to death. The worksheet enables students to work this out.

C: Will smoking take you out?

Since we know how much more likely smokers are to die of lung cancer than non-smokers, we can work out the chance that a death from lung cancer was due to smoking. The number we get is a probability, the risk that can be attributed to smoking; it is called the 'attributable risk'. (See Worksheet 6.)

There is a discussion of the concept, using some extreme cases, in the *Content epilogue*.

I explained the calculation of attributable risk using the example below. This is not easy and it is wise to take it rather slowly.

The attributable risk is the fraction (or percentage) of smokers' lung cancer deaths that are due to the fact that the person is a smoker. We assume that, if the person was not a smoker, they would have had the same chances of dying from lung cancer as non-smokers.

So, for heavy smokers, the relative risk is 17.0, which means that they have 17 times the death rate (due to lung cancer) as non-smokers. The increase due to smoking is then from 1 to 17.0, that is 16.0. So the attributable risk is the fraction 16.0 ÷ 17.0 or 94%.

The method we used was (17.0 – 1) ÷ 17.0.

In general terms the relative risk can be changed into an attributable fraction using the formula:

$$ar_s = \frac{rr_s - 1}{rr_s}$$

The graph on the right shows the relationship.

The answers to Worksheet 6 are given in the *Content epilogue*.

Attributable risk for smokers

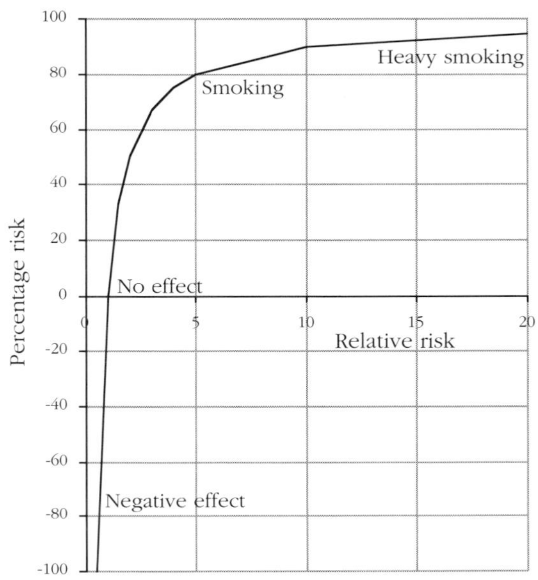

D: Teenage drinking

An issue related to smoking, that of teenage drinking, is addressed in Worksheet 7. The data comes from the 1990 national survey conducted by the Anti-Cancer Council of Victoria.

Content epilogue: background information for teachers

Cancer

Cancer is a term describing a large number of diseases with one common characteristic: they are the result of uncontrolled cell division (growth). They can arise in any part of the body. Cancer cells do not stay in the tissue in which they started; that is, they spread to parts of the body which are distant from the original site.

The Anti-Cancer Council is the best source of information and material for use in schools on this topic. Contact their office in your state.

Students will want to talk about personal situations, about the sufferings of people they know, and about why it happens. This sequence of lessons aims to highlight some of the causes less likely to be raised in public due to the influence of industries involved.

Cancer Issues Resource: An Activity Book for Post-Primary Teachers (from the Anti-Cancer Council, 1 Rathdowne Street, Carlton South, Victoria 3053) is a great help in finding a way to talk about such a disturbing topic with your class. This book also provides much information that can be useful for their background (and possibly yours). Refer to it, or other factual sources, for background information that will be part of the discussion.

In addition, try *The Tobacco Action Pack* (from the Victorian Smoking and Health Program, PO Box 888, Carlton South, Victoria 3053).

The meaning of attributable risk for smokers (see Worksheet 6)

The general definition of attributable risk for smokers (ar_s) is simple. (The relative risk for smokers is rr_s.)

$$ar_s = \frac{rr_s - 1}{rr_s}$$

To see more clearly what this means, imagine several extreme cases.

- If all the lung cancer deaths were of smokers, then $ar_s = 1$. The risk is 100%. This means, if you got lung cancer, it would be 100% certain that it is because you smoked. Put another way, 100% of lung-cancer deaths would be because of smoking.
- If smoking made no difference, that is, that there is no extra risk for smokers, then the rates of deaths for smokers and non-smokers will be equal, $rr_s = 1$, and the attributable risk is 0. If you get lung cancer, it is not because you smoke.
- If the death rate for smokers is twice the rate for non-smokers, $rr_s = 2$. Then the attributable risk is 0.5 or 50%. If you get lung cancer, it is 50% certain that it is because you smoke. Put another way, 50% of lung-cancer deaths are because of smoking.

Answers to Worksheet 6

		Relative risk of dying	Attributable risk
Lung cancer	Current smokers (worked example)	8.9	89%
	1 Ever smoked	6.9	86%
	2 Heavy smokers	17.0	94%
	3 Ex-smokers	4.3	77%
Chronic bronchitis	4 Smokers	5.3	81%
	5 Ex-smokers	2.4	58%
Heart disease (35–44 years)	6 Ever smoked	3.3	70%
	7 Current smokers	3.7	73%
	8 Heavy smokers	4.4	77%
	9 Ex-smokers	1.7	41%

Questionnaire design

Two issues become very important to a designer of the questionnaire.

- You must be very clear about just what it is that you want to ask, and the final form in which you would like to present the results. The questions will need to cover some personal data (such as age, sex, year level) if these are to be related to the answers about level of smoking and/or drinking.
- You must be very careful about the wording; bias and vagueness must be removed as much as possible so that the responses are valid. The questions must be easy to understand and easy to read; there should not be too many of them. Multiple choice answers (tick the box) are the simplest way to limit the responses to the questions; this will simplify the later handling of the data.

Worksheet 1: National survey of teenage smoking

The results of the national survey differ according to the ages of the people involved. Here is the data from the survey (the figures are percentages of the whole sample of males or females at each age), and below are graphs of it.

Teenage smoking 1990

Frequency of smoking	12 years M	12 years F	13 years M	13 years F	14 years M	14 years F	15 years M	15 years F	16 years M	16 years F	17 years M	17 years F
Never smoked	61	69	49	53	41	41	32	29	29	25	27	24
<10 ever	7	5	8	10	11	11	13	11	11	15	12	13
Weekly	6	5	11	13	17	20	22	29	25	28	24	28
Daily	0.5	0.5	3	2	6	6	10	11	12	11	12	11

Smoking: 12 years

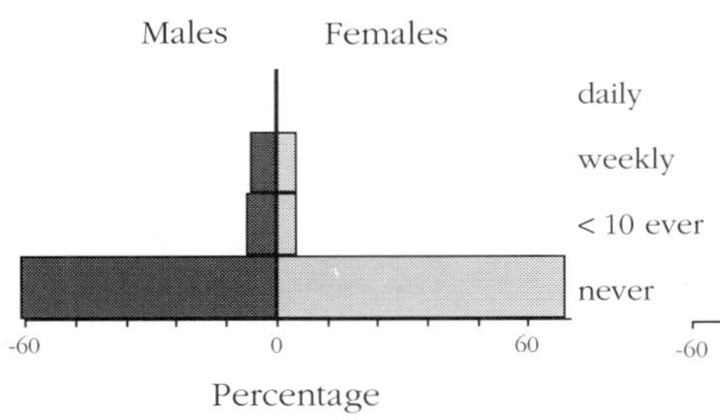

Smoking: 13 years

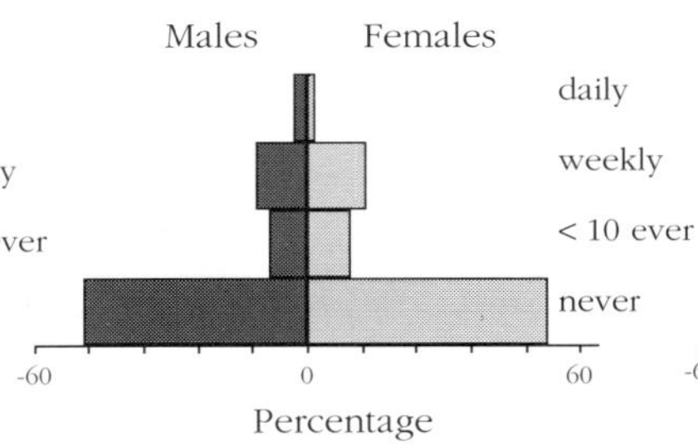

Smoking: 14 years

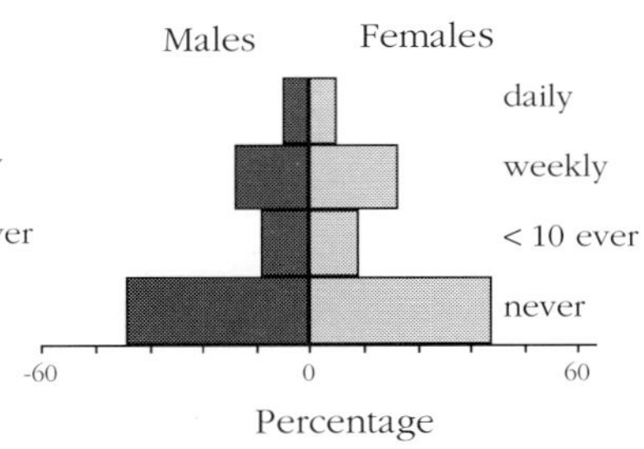

Smoking: 15 years

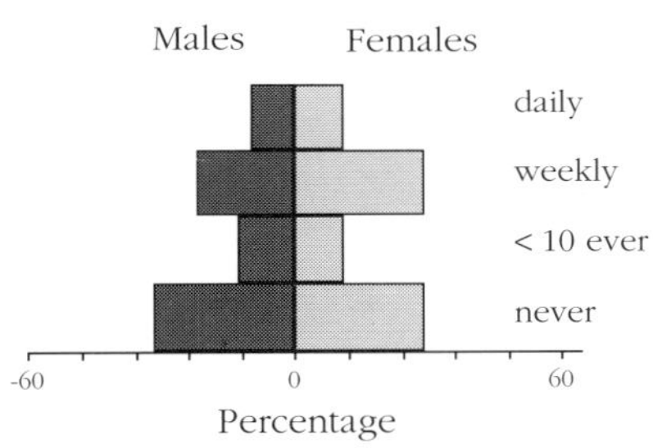

Smoking: 16 years

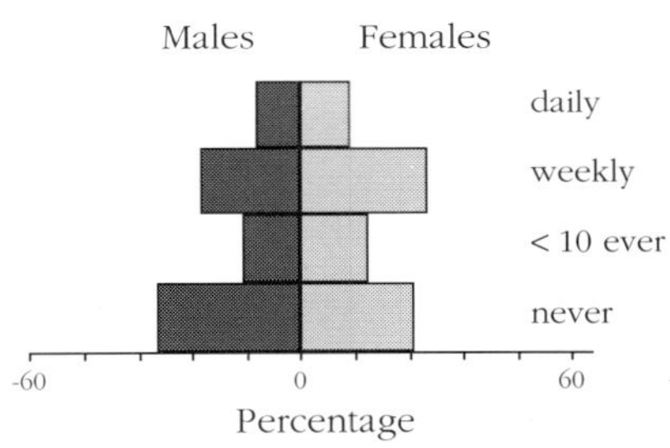

Smoking: 17 years

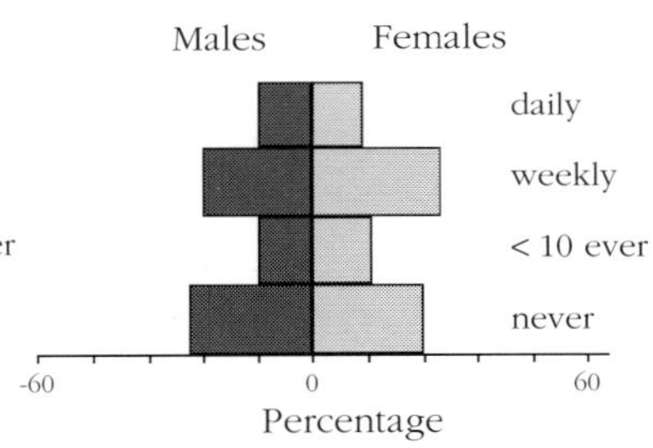

How much did they smoke?

At all ages boys smoked more than girls. At 17 years of age, those boys who smoked, smoked an average of 43 cigarettes per week, and those girls who smoked, smoked an average of 30 cigarettes per week. The most common brand in 1990 was Peter Jackson.

Discussion questions

1 Discuss what this means, then write about what it tells you.

2 How do the percentages change as the teenagers get older?

3 How do the figures for boys and girls compare?

The figures suggest there is a reduction in the percentages who smoke at ages 16 and 17.

4 Why might this be so? (Hint: the surveys were taken at school.)

5 Is this apparent reduction likely to be genuine? What do you think?

Worksheet 2: Lung cancer and tobacco imports

Trend graphs of lung cancer and tobacco imports are shown.

Is there a statistical relationship between tobacco imports, and thus cigarette consumption, and lung-cancer deaths?

Does this *prove* that cigarette smoking *causes* lung-cancer deaths?

Annual tobacco imports and average annual lung-cancer deaths

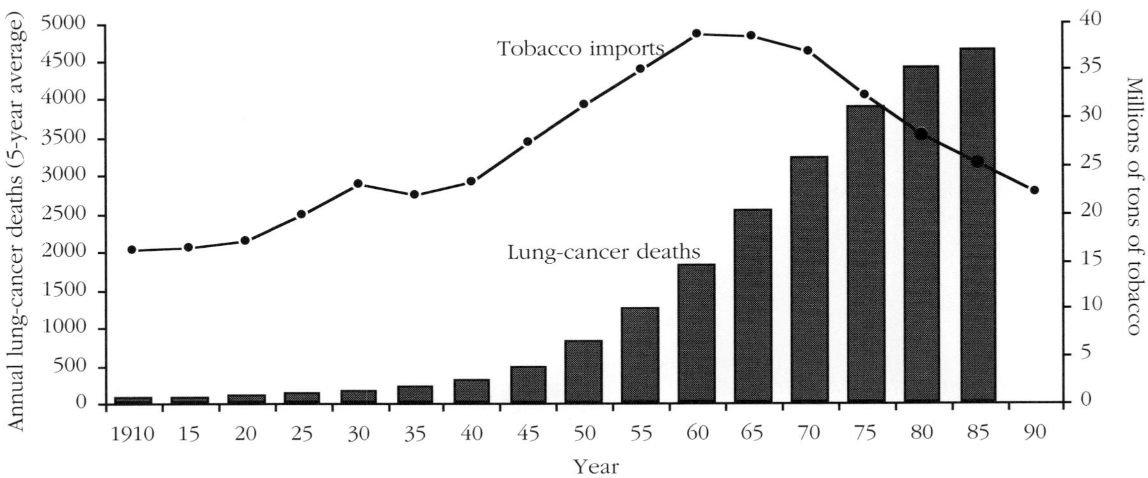

USA annual lung-cancer deaths at age 55–64

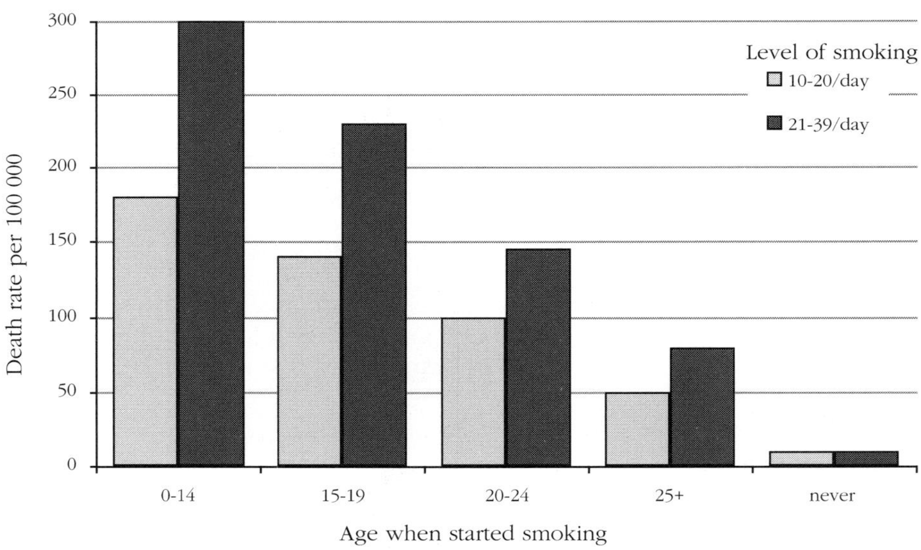

How is the death rate affected by the age at which the smoking habit is started?

How is it affected by the level of smoking?

Worksheet 3: Lung-cancer deaths in 1990

In Australia in 1986 it is estimated that 17 800 deaths, or 15.5% of total deaths, were due to use of tobacco. About three times as many males as females died from these causes.

The major causes of these deaths were heart disease, lung cancer and chronic bronchitis.

Lung-cancer deaths in 1990 — gender/age specific numbers and rates

Age group	0–1	2–14	15–24	25–44	45–54	55–64	65–74	75+	Total
Numbers of deaths									
Males	0	1	3	68	336	1217	1840	1436	4901
Females	0	1	0	59	130	376	591	507	1664
Populations (in 100 000s)									
Males	1.316	17.861	14.138	27.280	9.275	7.333	5.313	2.799	85.316
Females	1.248	16.991	13.537	26.755	8.810	7.257	6.211	4.730	85.538
Death rates (per 100 000)									
Males									
Females									

Source: *Causes of Death*, Australian Bureau of Statistics 1991

Lung-cancer death rates 1990

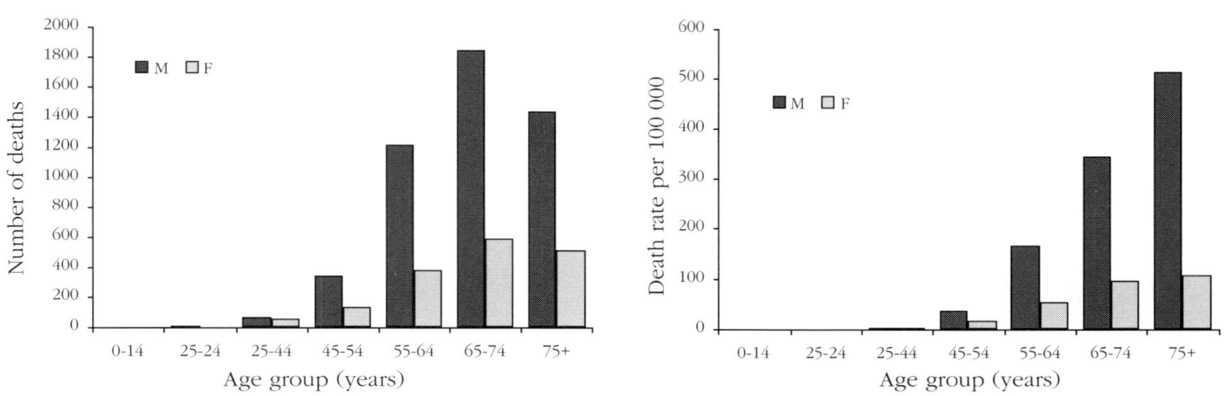

1 Which age groups suffer most from death by lung cancer?

2 How did men and women compare for lung-cancer deaths in 1990?
How does this depend on age?

3 Use the population and death data to calculate the death rates (to the nearest whole number).
Which is more helpful: deaths or death rates? Why?

It is estimated that about 60% of men and about 45% of women have been or still are smokers. In 1990 there were 6565 deaths due to lung cancer.

4 How many deaths is this per week? How many per day?

If you have ever smoked, and you die of lung cancer, the chance that you died because of your smoking habit is 86%.

5 How many men and women died in 1990 because they were smokers?

Worksheet 4

Deaths due to chronic bronchitis

Smoking is a major cause of chronic bronchitis. It is estimated that 81% of the deaths of smokers due to chronic bronchitis occur because of the smoking habit. For people who have given up smoking, 28% of males and 21% of females who die of chronic bronchitis do so because of their past smoking.

Chronic bronchitis death rates 1990

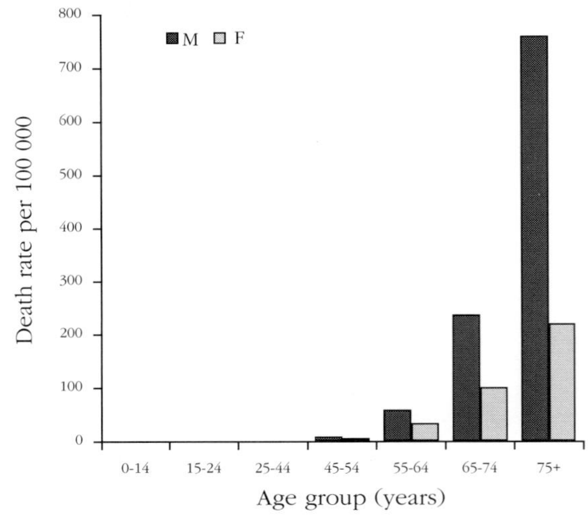

Deaths due to heart disease

Heart disease is a term that includes any diseases that relate to blood circulation. About a third of the deaths of Australians are due to some form of heart disease; it is the biggest killer of all. The three main diseases that are included are listed below.

Hypertension or high blood pressure causes damage to the heart, kidneys and eyes. It can lead to heart failure or to strain on the arteries of the brain giving rise to a stroke. Its causes are related to lifestyle (stress and inactivity) and also to diet (salt and alcohol).

Arteriosclerosis or hardening of the arteries leads to heart attacks, strokes and kidney failure. Raised blood cholesterol is a major cause of this condition; other major causes include smoking and diabetes.

Coronary heart disease is the major cause of heart attack. Except for those with diabetes, males are about five times more likely to suffer heart attacks than females. The cause may be hardening of the arteries, but may also be due to a small blood clot preventing flow and leading to a lack of oxygen.

The National Heart Foundation is the best source of information and resources suitable for schools. Contact their office in your State.

Heart disease death rates 1990

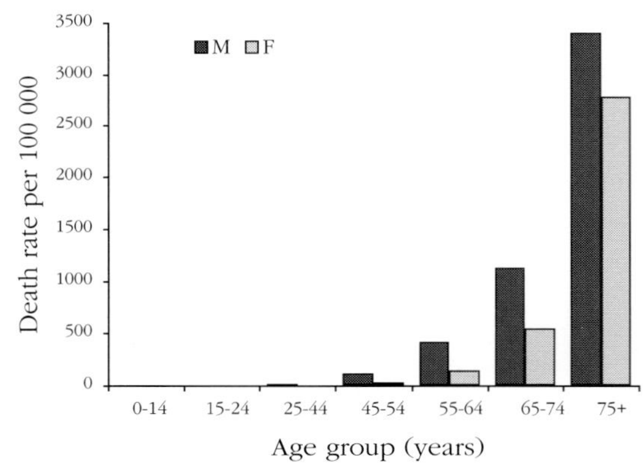

Relative risks

Smokers who smoke more than one pack per day are 4.4 times more likely than non-smokers to die of heart disease between the ages of 35 and 44. For those who smoke less than a pack per day the ratio is 2.7, and it drops to 1.7 for those who give up smoking.

For heart disease deaths of people over 70, the ratios are 1.7 (over one pack per day), 1.5 (under one pack per day) and 1.2 (ex-smokers).

Compare the vertical scales on the death rate graphs for lung cancer, chronic bronchitis and heart disease

How is heart disease related to age and level of smoking as a cause of death?

How is heart disease related to age and gender as a cause of death?

Write your own statement interpreting the graphs above, to describe how heart disease strikes males and females at different ages.

Worksheet 5: Comparing smokers and non-smokers

The cause of a person's death is listed on their death certificate. In 1990, 6565 people died of lung cancer. Surveys have been used to determine the fractions in the community who are or are not smokers, and how often and how much they smoked. From these numbers we can compare the rates at which smokers and non-smokers die of lung cancer and calculate the relative risk between them. Here is how it is done.

There were approximately 8 531 600 men alive on 30 June 1990. During 1990, 4901 men died of lung cancer.

Group	Number	Number of deaths	Rate per 100 000
Total male population (1990)	8 531 600		
Males who have ever smoked (60.6%)	5 170 150	4 479	86.6
[The death rate per 100 000 people is (4 479 ÷ 5 170 150) × 100 000 = 86.6]			
Males who have never smoked (39.4%)	3 361 450	422	12.6

Relative risk = 86.6 ÷ 12.6 = 6.9

In other words a male who has ever smoked is 6.9 times more likely to die of lung cancer than someone who has never smoked.

1 Now repeat the above table for females. There were approximately 8 553 800 women alive on 30 June 1990. During 1990, 1664 women died of lung cancer.

Group	Number	Number of deaths	Rate per 100 000
Total female population (1990)			
Females who have ever smoked (45%)		1 410	
Females who have never smoked (55%)		254	

Relative risk =

2 Repeat the calculations for the following groups.

Remember to get the relative risk you compare the rate per 100 000 with the non-smoking rate (which is 12.6 for males and 5.4 for females).

Group	Number	Number of deaths	Rate per 100 000	Relative risk
Male heavy smokers (14.1%)		2 638		
Female heavy smokers (10.2%)		775		
Males who were current smokers (32.9%)		3 207		
Females who were current smokers (28.5 %)		1 129		
Males who had given up smoking (27.7%)		1 309		
Females who had given up smoking (16.5%)		317		

3 Does giving up smoking reduce your chance of dying of lung cancer?

Worksheet 6: Will smoking take you out?

The relative risks you calculated on Worksheet 5 should have given you this picture. The central ellipse shows the smokers in the population. The small circles are those who die of lung cancer due to different levels of smoking using averages of male and female calculations.

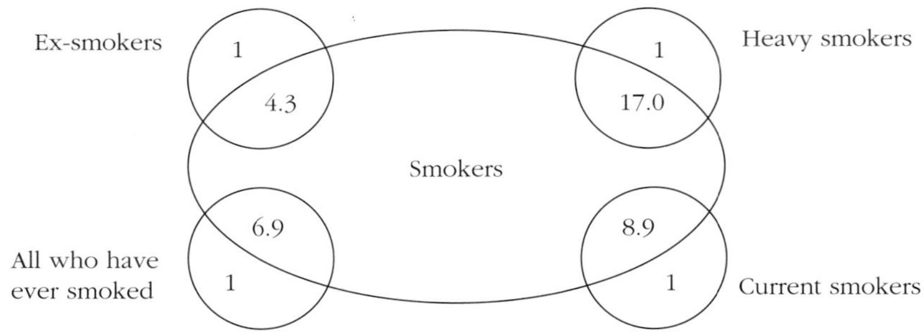

So, a smoker can die of lung cancer, smoking being a major cause of this disease. A person who is still smoking has about ten times the chance of dying from lung cancer than a person who is a non-smoker.

Now if a smoker dies of lung cancer, it could be from other causes, such as those (for example passive smoking) which cause the non-smokers to die from lung cancer. But compared to non-smokers, there is an increase in the chance of dying from lung cancer of nine times the non-smoking rate.

So this increase in the rate is entirely due to the habit of smoking. Put another way, nine out of ten lung cancer deaths of smokers are due to them being smokers; the other one out of ten is what would have occurred if they were non-smokers. So we can say that 90% of the deaths of smokers who die of lung cancer are attributable to smoking.

For any 'current smoker', we can never be quite clear that a lung cancer death is caused by smoking, but we can say that there is a chance of 88.8% that a particular death is due to smoking. This is called the *attributable risk*. The method involved is simply to subtract 1 from the relative risk, and divide the result by the relative risk. In this example $(8.9 - 1) \div 8.9 = 0.8876 = 88.8\%$.

Use the same method to find the attributable risks of lung cancer for the other groups whose relative risks are shown in the diagram above:

1 Lung cancer: all those who have ever been smokers

2 Lung cancer: current smokers

3 Lung cancer: ex-smokers

Smokers can also die of chronic bronchitis or heart disease. The relative risks are given below. Work out the attributable risks for these groups.

4 Chronic bronchitis: all those who have ever been smokers (relative risk 5.3)

5 Chronic bronchitis: ex-smokers (relative risk 2.4)

6 Heart disease between 35 and 44: all those who have ever been smokers (relative risk 3.3)

7 Heart disease between 35 and 44: current smokers (relative risk 3.7)

8 Heart disease between 35 and 44: heavy smokers (relative risk 4.4)

9 Heart disease between 35 and 44: ex-smokers (relative risk 1.7)

Worksheet 7: Smoking and drinking

In the following table the numbers show the percentage of the sample of males and females at each age who drank to the extent listed.

Teenage drinking 1990

Frequency of drinking	12 years		13 years		14 years		15 years		16 years		17 years	
	M	F	M	F	M	F	M	F	M	F	M	F
Never	22	24	18	17	11	11	7	6	6	4	5	3
<10 ever	12	7	18	16	22	21	21	23	15	22	14	17
Monthly	20	14	29	24	39	36	52	50	60	61	67	66
Weekly	13	8	20	17	28	25	38	34	44	43	51	46

Frequency of drinking

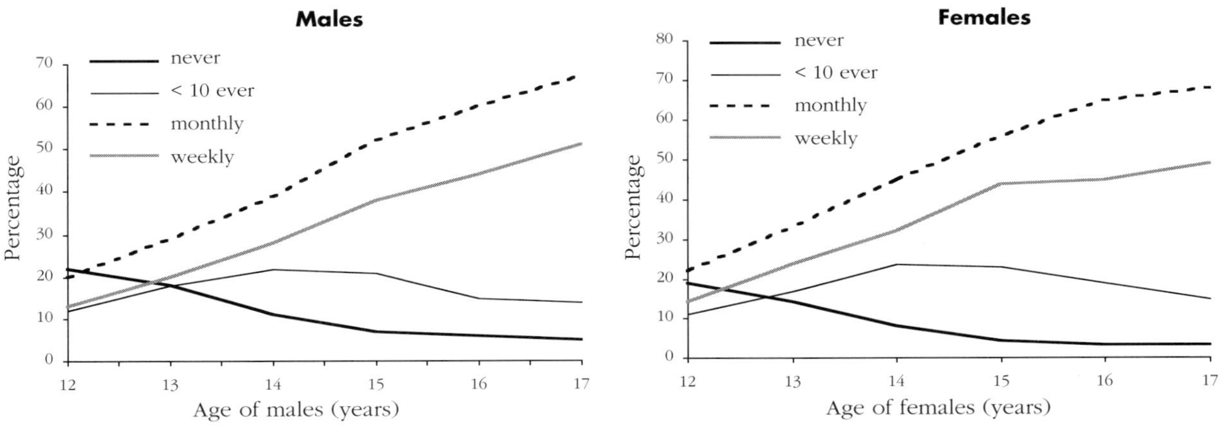

At all ages boys drink more than girls, up to nine drinks per week at age 17. Most drinking takes place at weekends.

Relationships: smoking and drinking

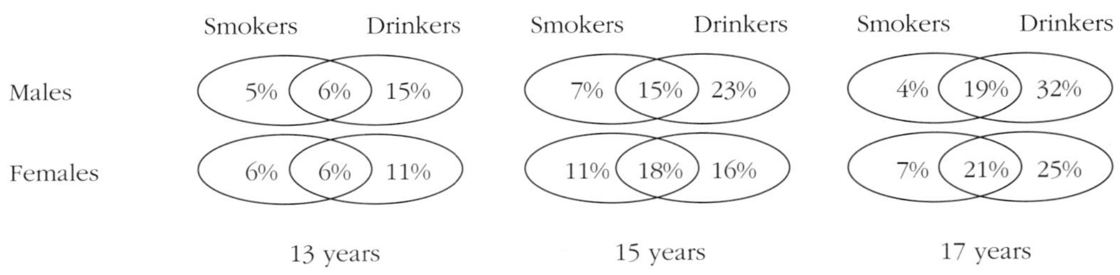

1 How would you describe this situation? What 'story' is told by these figures?

2 What relationships can you detect between smoking and drinking?

3 Teenage smoking leads to lung-cancer deaths in later years — see Worksheet 2. Which habit, drinking or smoking, seems to be more dangerous to public health? Give your reasons.

Sunsmart?

Years 9 to 12

Advertisements tell us that two out of three Australians will get skin cancer in their lifetime. Just what are the facts about skin cancer? How do Australians compare with the rest of the world? Cancers are a common cause of death; how have the rates changed over the years? Which age groups are mainly affected? The social issue is revealed through a study of the relevant data. *Dying for a smoke* is a closely related investigation looking at the effects of smoking, particularly on lung cancer.

Features

Teaching and learning

- Exposes a sensitive social issue through real data
- A guided investigation
- Class discussion and use of outside experts
- Group discussion and reporting

Major mathematical content

- Interpreting secondhand data in the form of graphs and tables
- Constructing arguments using data

Mathematical content is further discussed in the *Content epilogue*.

Preparation and equipment

- The data and graphs are on the worksheets. For background, contact the Anti-Cancer Council (or equivalent) in your State. You might call on a doctor or nurse experienced in this area
- The video referred to in stage 1 has been screened over the summers of 1991 and 1992 — your students may remember it

Expected timing

- This investigation could range from one or two class sessions, up to several weeks, particularly if it is linked with *Dying for a smoke*

A teacher's story

I had felt the need to put some real, non-trivial data into the mathematics program for some time. Cancer seemed to me to be a most important issue, and of real concern to many students. With the cooperation of the Anti-Cancer Council, I extracted the material I wanted and led the class through it. I started with skin cancer as it has a high profile due to advertising, and is less threatening to talk about than many other, more deadly, cancers.

Students need some time to discuss the meaning of data and graphs. One of the best ways to allow this is to create small discussion groups that will report to the whole class after about 10 minutes. In this way each group gets to think for themselves, and to present their findings. Because they will mostly agree on the interpretation, the whole group quickly reaches a consensus.

My method of getting a good discussion going involves two parts: group discussion and reporting.

Group discussion: I like to pose the questions clearly, even writing them on the board. Then I set a reasonable, but definite, time limit and break the class into groups for discussion. One member of each group takes the role of leader, keeping them on task, and making sure all are involved. A record keeper takes notes to be used by another group member who reports the discussion to the class. This reporter summarises the findings by making several succinct statements agreed to by the group.

Reporting: I ask the reporter from each group to give us the single major point arising from their discussion. If this has been covered by a previous report, a new point is added. The reports will overlap, which forms the consensus. My role is to ask leading questions that will help focus the responses on the main task. Where there are disagreements, the protagonists can argue their cases publicly, and if needed the class takes a vote.

Comments

The class was amazed at the close relationship between latitude and amount of skin cancer.

They became very skilled at summarising lots of data and forming hypotheses about the relationships that were shown.

We found that the conclusions we could not help arriving at were rather depressing, so we left the investigation for a few weeks and returned to it later.

Summary of the stages in this investigation

1 Introduction to skin cancer

2 Skin cancer incidence rates and age

3 International comparisons: latitude, race and gender

4 Trends in cancer deaths

5 Some other cancers

1 Introduction to skin cancer

I started the discussion by recalling the advertisement 'Leave your hat on', which was screened frequently in Victoria over summers of 1991 and 1992. It made this claim:

> Two out of three Australians will get skin cancer in their lifetime.

This is a statistical fact, but there are several forms of skin cancer, many less harmful than melanoma — a form of skin cancer able to cause death by spreading to other parts of the body.

Students were aware that there is an attempt to get them to cover up instead of sunbathing, and to use a sun-block cream whenever they go out into the sun.

I asked them to look at the newspaper material on Worksheet 1 while I read it aloud. It paints a picture of a disease growing exponentially at ever increasing rates. We explored these relationships briefly with our calculators.

Doubling each 15 years is equivalent to about 5% growth each year ($1.05^{15} = 2.08$), but doubling in 7 years is equivalent to about 10% growth each year ($1.1^7 = 1.95$).

The lifetime incidence of melanoma (about 1 in 25) can be contrasted with the incidence of any skin cancer (about 2 in 3).

We quickly moved into discussing cancers in general. I found that students were quite ignorant about the disease and because of its relationship to death were initially uncomfortable about discussing it. However, once we had been at it for a few minutes they realised that this discussion was of real personal value to them, and joined in enthusiastically.

The discussion needs to be handled with care and sensitivity. Because cancer figures significantly on the list of causes of death, this increasingly prevalent disease needs to be treated objectively. Be careful not to press students who are reluctant to become involved; they will be listening to the dialogue.

I found that my students started to volunteer information about family and friends who had suffered from various cancers. We found that the effect of cancer was much more widespread than we had ever realised. It seems that it is just not talked about, so we don't realise how devastating it is.

2 Skin cancer incidence rates and age

The reasons for the current concern about skin cancer can be explored by comparing the incidence of melanoma in Queensland (the highest in the world) to that in Victoria. Data relating the rates to gender and age are on Worksheet 1. There are some questions to help guide the discussion of this material.

I divided the class into groups and asked each group to prepare a report for the class on the implications of this data, using the questions on the sheet to guide them.

More Queenslanders get melanomas than Victorians. People sunbake more in Queensland.

Older people get melanomas; young people are less likely to get them.

Latitude and gender are important factors, but age is the most significant factor. Why?

Skin cancers take many years to develop.

Maybe the 1950s ideal of a bronzed skin was a mistake that is costing many people their lives.

The older you are, the more you have been exposed to any harmful effects of sunlight.

Will the hole in the ozone layer make this worse?

The effects on skin cancer do not show up for several decades; then it is far too late to do anything.

3 International comparisons: latitude, race and gender

We had started to see the effects of being nearer the equator in comparing Queensland and Victorian rates. The international picture added a whole new dimension!

Once again I used group discussion (just 10 minutes) followed by brief reports.

The graphs on Worksheet 2 show how racial group makes quite a difference to the level of incidence of melanoma. Note the effect of both latitude and racial grouping on the rates. Queensland leads the rates and other parts of Australasia come above the rest of the world.

We looked at the way melanoma incidence is related to latitude for white-skinned people in the northern hemisphere. I arranged for a large wall map (from the Geography department) and we looked at the locations of the countries whose data were given.

What is the effect of where you live for white Australians?

What relationship can you suggest from this data?

The data for non-white-skinned people are shown on the third graph. What can you say about these rates compared with the rates for white-skinned people?

On Worksheet 2 is a scatter-plot of the male and female rates, to enable a focus on gender. The scatter-plot has a line at 45° to show the cases where the two rates are equal. So it is fairly easy to see where the male rates exceed the female rates, or female rates exceed the male rates.

Are males or females more likely to get melanomas?

Worksheet 3 is provided for graphical-interpretation practice, and as a reminder about prevention.

4 Trends in cancer deaths

The graph on Worksheet 4 shows the death rates due to the most common types of cancer over the course of this century; the rapid rise in lung cancer is very clear, as is the fall in stomach cancer. (There is more about lung cancer in *Dying for a smoke*.)

I used this material to stimulate group discussion. Again, sensitivity is required, particularly if someone's relatives or friends are currently suffering or have recently died.

The table on Worksheet 4 gives the death rates for several common cancers in averages of 5-year groups from 1910. Students are invited to graph the data and see how the rise in lung-cancer rates has mostly caused the increasing total.

5 Some other cancers

The table on Worksheet 5 gives the number of deaths in 1990 due to all forms of cancer. Some analysis questions are suggested.

Students can calculate the rates for each gender and age group. The gender/age specific rates are found by dividing the number of deaths for that age and gender by the total population in that group, and multiplying by 100 000.

Two further graphs are presented for analysis:

- breast-cancer death rates for different age groups of women;
- colon-cancer death rates for different age and gender groups.

Students should note the different vertical scales on these graphs, and try to form an idea of their relative impact.

The data on Worksheet 5 come from the 1991 annual ABS publication *Causes of Death*. Contact the ABS in your State capital for the most recent copy.

Breast cancer is the top cause of cancer deaths for women. The material in the *Content epilogue*, 'What causes breast cancer?', is from the Anti-Cancer Council of Victoria, and is printed with their permission. It appears in their free brochure *Breast Cancer: A Guide for Women*.

Colon cancer is the third most common cause of death by cancer for both men and women.

Content epilogue: background information for teachers

Exponential growth and doubling time

The newspaper article on Worksheet 1 mentions the time in which the incidence of melanoma doubled. This is a signal to mathematicians that the disease is increasing exponentially. If it continues to double at the same rate for some time it is following a 'compound interest' type of formula:

$$y = a\,b^x \qquad \text{where } b \text{ is the base and } x \text{ the number of years}$$

By experimenting with a calculator we can find the value of x (number of years) that allows the value of y to double. It is, of course, the value of x that makes $b^x = 2$. For example, to find the rate (b) for doubling in 15 years ($x = 15$), we could just try different values of b until we hit on one that works. This is a good method for students to use.

The more advanced method takes the log of both sides:

Thus $$b^{15} = 2$$

so $$15 \log b = \log 2$$

or $$\log b = \frac{\log 2}{15}$$

Using the calculator and taking the inverse log we get:

$$b = 1.047$$

This means that the increase each year is 4.7%.

The change in doubling time from 15 to 7 years corresponds to a change in annual rate from 5% to 10%. The relationship between the rate (r%) and the doubling time (t years) is inverse: $rt = 70$ (approx.).

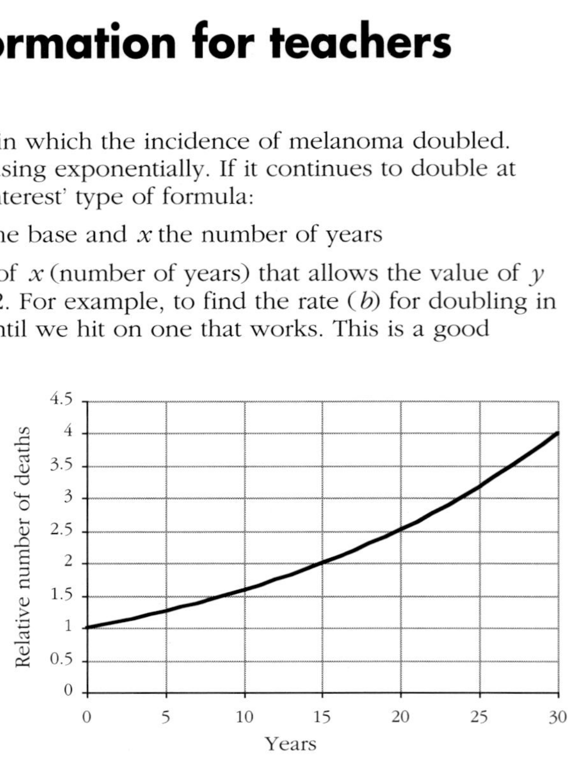

Interpreting graphs

The easiest way to get an idea of what is happening in a table is to graph it, even if only in a rough sketch. For example, the graphs on Worksheet 1 show four trends. Each trend is interesting in itself. However, the most important ideas are the differences between the trends. (See also the table on the right.)

- Are males more likely to get melanoma than females? At what ages? Why might this be?
- Why are Queenslanders more likely to get melanoma than Victorians?

The first two graphs on Worksheet 2 show the approximate latitude of cities, States and countries, and the melanoma death rates for white males and females. Only when the white-skinned populations are separated do we see a clear relationship to latitude. (Latitude is here represented horizontally, unlike the map convention.) Other racial groups rarely get melanoma so the death rates are much lower in the third graph for non-whites.

Incidence of melanoma (per 100 000)

Queensland	
Males	40.4
Females	41.9
Victoria	
Males	22.6
Females	24.1

Note that these overall rates reflect the gender imbalance in the population, particularly in later years.

Worksheet 4 uses graphs and tabled data to show trends in death rates. These rates are averaged over five-year intervals. It turns out that, if lung cancer rates are removed, the total death rate from cancer has remained stable over many years.

The table on Worksheet 5 shows the number of deaths in 1990 from cancer for gender age groups. The deaths are divided by the total population in those gender age groups and multiplied by 100 000 to get the rates (per 100 000 people). That is why they are called gender/age specific numbers and rates. The graphs show death rates for two common forms of cancer.

What causes breast cancer?

The cause of breast cancer is unknown. However, it is unlikely that there is one single cause. A number of factors, some known and many unknown, probably work together to trigger its growth.

Some myths corrected:
- A bump or a blow to your breast will not cause breast cancer.
- Breast cancer cannot be caught from someone who has it.
- Breast feeding does not cause breast cancer, nor protect you from it.

Breast cancer usually occurs in women over 40 and all women are at risk as they grow older. Although breast cancer does occur under 40, increasing age is the main factor influencing the risk of getting breast cancer.

The factors listed below also seem to put women at slightly higher risk, but they explain only a small number of breast cancer cases:
- Women with a strong family history of breast cancer have an increased chance of getting breast cancer.
- Women who have already had breast cancer have a slightly higher chance of a new breast cancer developing in their other breast.

Other factors which possibly influence the development of breast cancer include:
- a diet high in animal fat;
- not having children or having the first at a late age;
- some unusual types of benign breast disease.

Epidemiology and cancer

Epidemiology is the statistical examination of the relationships between what people do and the diseases they get. In the case of cancer studies, effort has concentrated on exposure to foods and other possible 'causes', such as asbestos, cigarette smoking, etc.

One recent summary of 137 such studies relating the incidence of different cancers to the diets of the populations concerned was summarised by the authors in this way.

'It is concluded that consumption of higher levels of vegetables and fruit is associated consistently, although not universally, with a reduced risk of cancer at most sites... The association exists for a wide variety of vegetables and fruit with some suggestion that raw forms are associated most consistently with lower risk.'

Worksheet 1: Melanomas in Australia

On 3 January 1992, Julian Cribb, science and technology writer for *The Australian*,wrote this:

> Doctors believe parents may be more to blame than previously thought for the soaring incidence of the deadly skin cancer melanoma, which will kill one Australian in every hundred.
>
> Research by Sydney University scientists has shown that the melanoma epidemic sweeping the country may, in part, result from genetic defects inherited from parents — as well as parents allowing children to become badly sunburned when young.
>
> The incidence of melanoma had doubled in the past seven years, representing a serious acceleration in the rate of the disease, Associate Professor Richard Kefford, who heads the world's largest human melanoma genetics group, said.
>
> 'It's absolutely frightening. Previously, the rate of melanoma doubled only once in every 15 years. The disease seems to have entered an incredible phase of growth.'
>
> 'Melanoma accounts for about one tenth of all skin cancers, but because of its resistance to treatment is easily the most lethal, claiming as its victims a quarter of all cases', he said.
>
> It is now the second most common form of cancer in both sexes — and is exceeded only by lung cancer in men and breast cancer in women.
>
> The lifetime risk for an Australian of suffering from a melanoma is now about one in 25. At present a quarter of those who develop it die as a result — making the risk of a fatality about one in 100.

Incidence rates of melanoma-type skin cancers in Queensland and Victoria

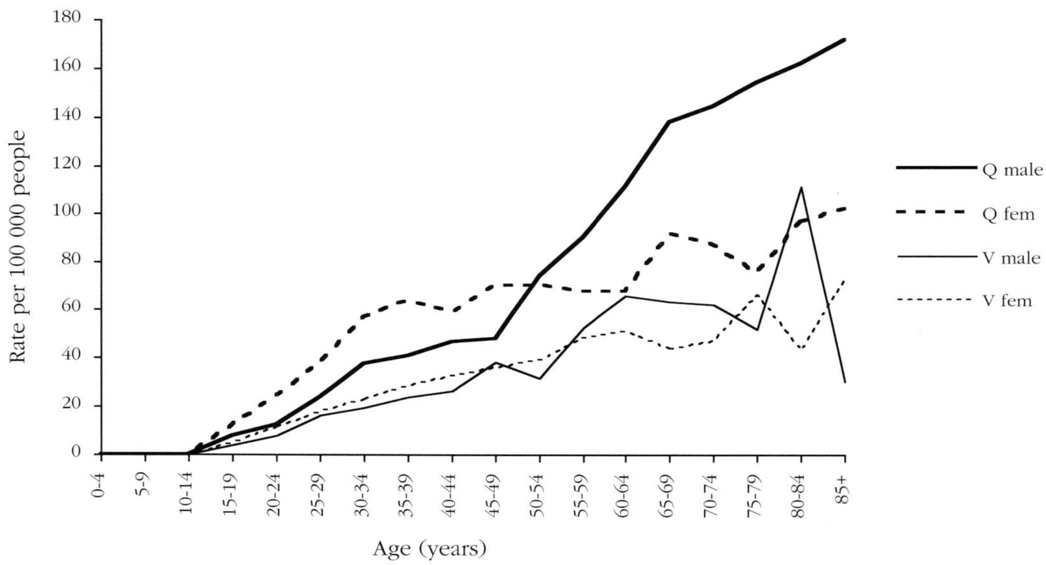

1 Discuss and write about the meaning of the graphs.

2 What are the factors that seem to lead to greater numbers of people with melanomas?

3 Why are the two States so different? Can you explain this?

4 Is there any clear relationship to age? Can you explain this?

5 Is there a relationship to gender? Can you explain this?

The unusually high value for Victorian males aged between 80 and 84 is due to the very low number of males still alive at that age. The rate arises from a total of seven deaths in the year (1987).

Worksheet 2: Melanoma death rates

The following graphs show international comparisons of melanoma death rates.

South latitude melanoma white death rates

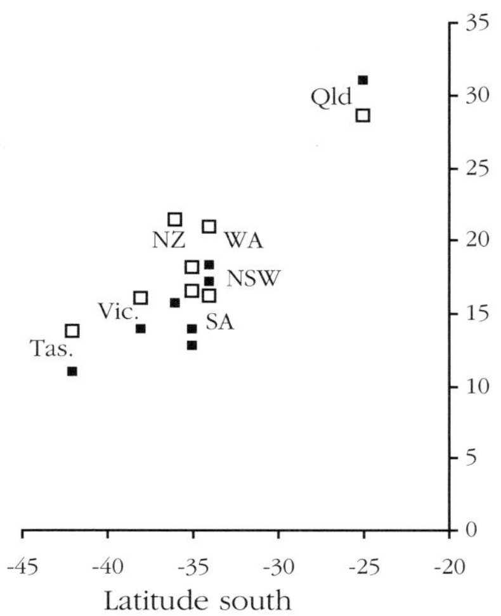

North latitude melanoma white death rates

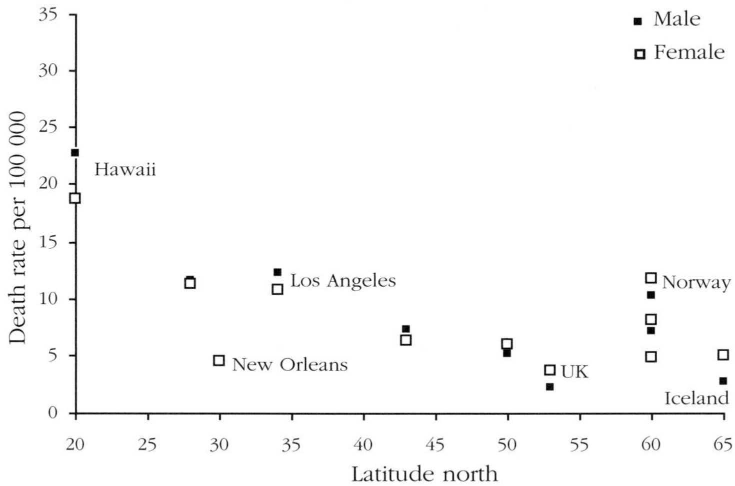

Melanoma non-white death rates

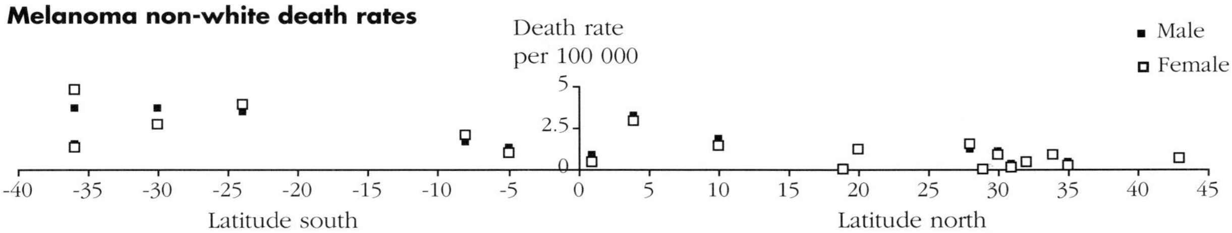

1 Discuss and write about the meaning of these graphs.

2 How does latitude seem to affect the melanoma death rate? Why might that be?

3 How does having white skin affect your chances of getting melanoma?

4 White Queenslanders have the highest melanoma death rate in the world. Why might that be?

Melanoma and gender

Here is a graph of the male and female melanoma death rates for the places in the graphs above. (Rates are in deaths per 100 000 per year.) This time only the male and female rates are plotted.

5 Why are there more dots near the origin? What does this mean?

6 Which places have the dots far from the origin? What does this mean?

7 Are there more dots above or below the line? What does this mean?

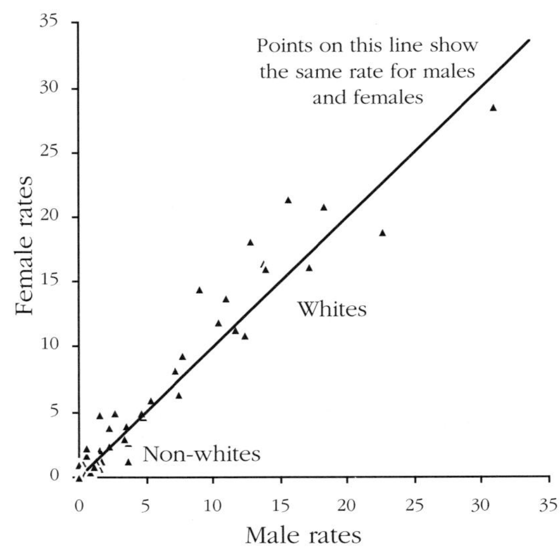

Worksheet 3: Staying out of the sun

The graphs below show how the total solar ultraviolet radiation and the part that causes sunburn (called UVB) vary. The graph on the left shows the variation at different times of the day, for different times of the year. The graph on the right shows how the maximum values of the total and the UVB part vary over different months of the year.

1 In 1987, what was the strength of the maximum radiation? When did it happen?
2 What was the strength of the UVB part on that day? Estimate the fraction of the total that was the UVB part.
3 What fraction of the January values are the July values?
4 Describe in words how the radiation changes over the year. Why does this happen?
5 Why is it important to 'slip under a tree between 11 and 3'?

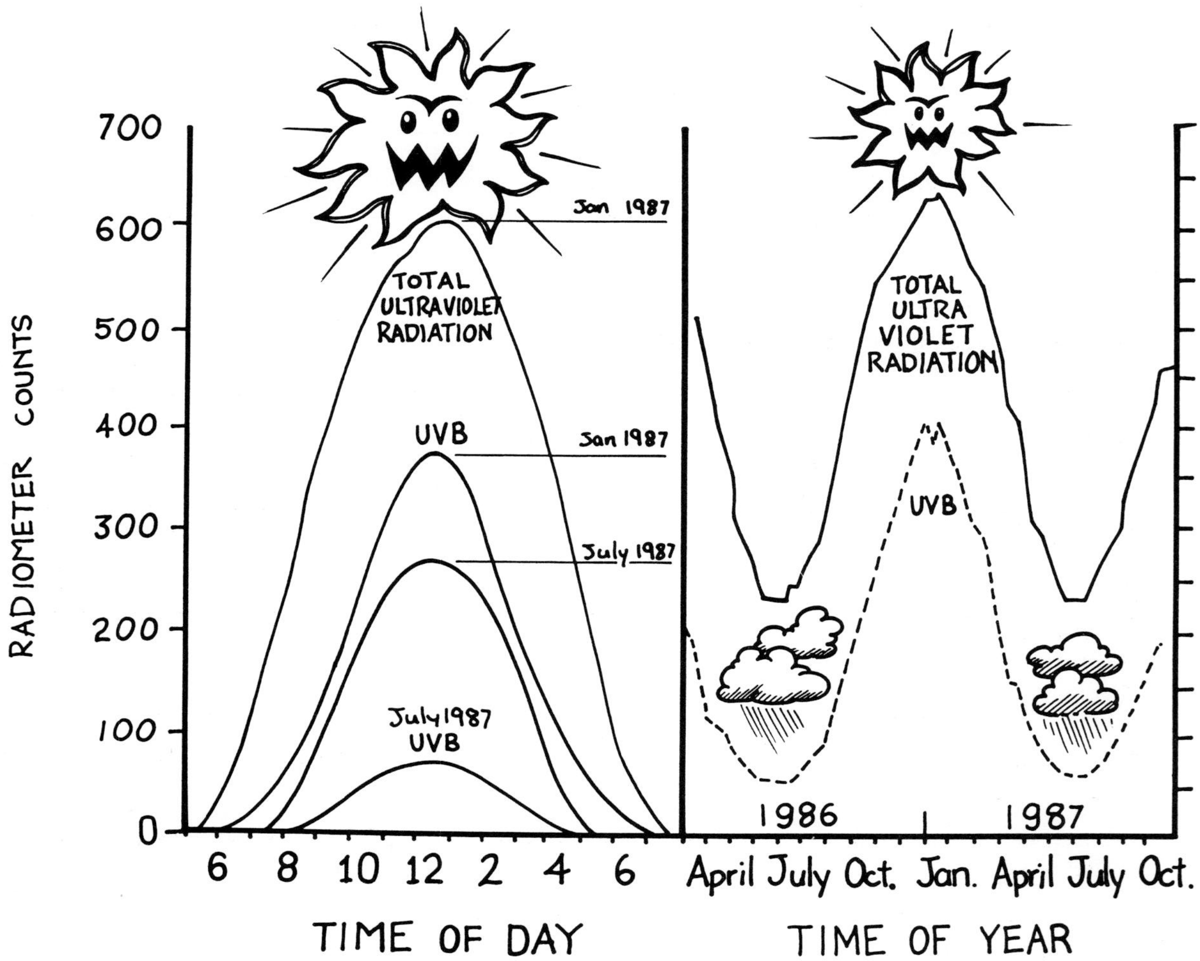

Worksheet 4: Trends over time

Trend graph of several common cancers

The graph data is taken from Holman, C.D.J. and Hatton, W.M. *Cancer Mortality Trends in Australia 1910–1984* (Health Department of WA 1987), and from the Anti-Cancer Council of Victoria (unpublished data).

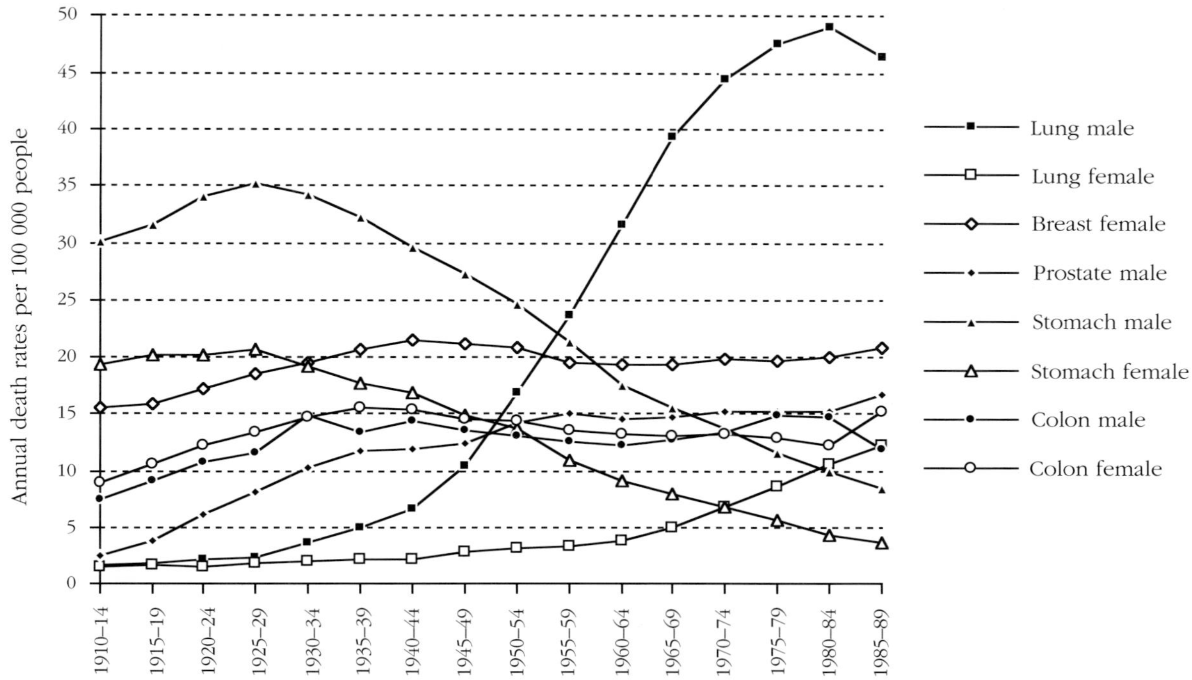

1 For each gender, what were the most common cancers in 1940–44?

2 For each gender, what were the most common cancers in 1985–89? How have these changed since 1944?

3 Discuss and write the story that this graph is telling.

Total cancer death rates, 1910–1989

Start of 5-year group	1910	15	20	25	30	35	40	45	50	55	60	65	70	75	80	85
Cancer death rates																
Males	109	115	121	120	118	119	118	120	126	133	135	149	157	161	164	164
Females	109	108	113	111	107	109	110	107	105	101	99	99	101	99	101	102
Lung-cancer death rates																
Males	2	2	2	2	4	5	7	11	17	24	32	39	44	48	49	46
Females	2	2	2	2	2	2	2	3	3	3	4	5	7	9	11	12
Total minus lung rates																
Males																
Females																

4 The death rates for all cancers have been increasing for some time. Draw a graph to show this. Write about the graph.

5 Lung cancer has been increasing rapidly for about the same time. Include its rates on your first graph.

6 Calculate the rates for total cancers leaving out the lung cancers. Plot this difference on your graph also. Comment.

Worksheet 5: Cancer and age

Cancer deaths in 1990

Gender/age specific numbers and rates

Age	0–14	15–24	25–44	45–54	55–64	65–74	75+	Total
Numbers								
Males	79	89	684	1251	3552	5718	5957	17 311
Females	56	54	793	1235	2356	3669	4990	13 153
Populations (in 100 000s)								
Males	19.177	14.138	27.280	9.275	7.333	5.313	2.799	85.316
Females	18.240	13.537	26.755	8.810	7.257	6.211	4.730	85.538
Rates per 100 000 people								
Males								
Females								

1 Which groups suffer most from cancer deaths? Draw a graph to show the trend.

2 Divide each group's deaths by its population (in 100 000s). This gives the death rate per 100 000 people. Draw another graph to show this trend.

3 Discuss and write a statement about the differences between the male and female death rates.

4 Which gives the most useful information: numbers or rates?

Breast cancer death rates in 1990

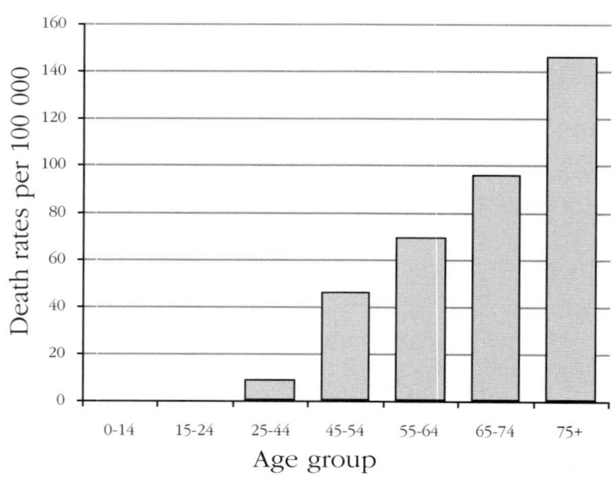

Colon cancer death rates in 1990

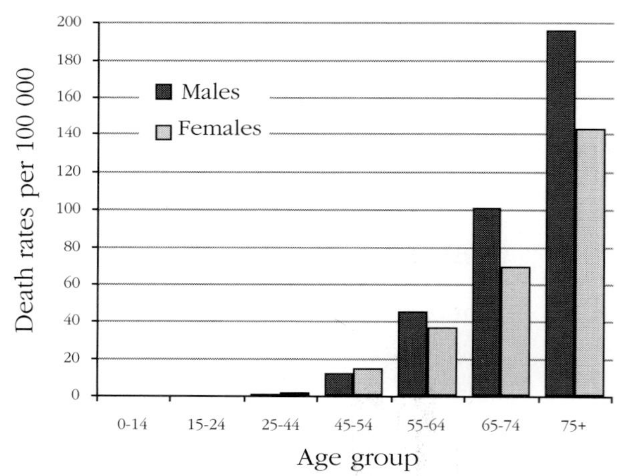

5 Which seems to be more common: breast or colon cancer?

6 The rates are quite different from one age group to another. Why might this be so?

7 Although most of the male colon cancer rates are above the corresponding female rates, they still have the same total rate of 18 deaths per 100 000 of total population. How can this be?

7 Using databases

These four investigations are intended for use with a computer database program; for example, Fredbase (public domain) for Apple II, Excel (or Works) for IBM or Macintosh, Graph for BBC, and so on. The investigations can be adapted for use without such a program, but they are written assuming a suitable program and access to at least one computer. However, they do not assume that you know anything about how to use the program. The first, *Database debut*, is a gentle introduction for both students and their teacher.

A database is an electronic device for manipulating data and for doing all the displays and calculations that you need. Simple databases are basically just tabulated data, but a good program allows you to manipulate the data just as you want. Consequently, most database programs will allow sorting, selection of subsets, and combinations of subsets. Only a few will create a good variety of graphs from the data, and only the best will draw histograms (or even rarer, box plots and stem-and-leaf plots).

The best time for students to learn how to use a database seems to be early rather than late, and the best method is to watch someone else who knows. This method is employed in each of the investigations that follows.

Database debut requires students to gather data by means of a questionnaire and to type the responses into the program. This is very much personal data, in the manner of chapter 5. However, the data disk has a set of data obtained in response to a similar set of questions. So the students can compare their personal data with this real, but secondhand, data. (To use this data in your own computer database program, you will need to follow its instructions for inputting data in 'text' or 'ASCII' form.)

A similar division between personal and provided data exists for *Cemetery maths* and *Pocket money*. For *Weather patterns*, some data is provided in 'hard copy' (printed page) but much more extensive data are in database form, on disk. Students could enjoy the experience of collecting some personal weather data (temperature and daily rainfall), but this is not required as part of the investigation.

Further practice of database programs can be obtained using the disk for *Exploring real data*, which is part of this *Problem solving in mathematics: Chance and data* package.

From *A National Statement on Mathematics for Australian Schools*

(Overview) Technological changes have influenced the techniques of data collection, retrieval, manipulation, analysis and communication, and have increased our capacity to pursue investigations with large quantities of real and simulated numerical data. The school curriculum should reflect these changes and students should have the opportunity:

- to use computing technology to enable them to handle data drawn from everyday contexts;
- to engage in extended, practical investigations;
- to consider the social implications of the impact of technological change on the collection and handling of data.

(Band C) It is appropriate to have students calculate summary statistics and draw graphs manually for some small data sets, in order to learn the techniques and so deal with situations where very little by way of data manipulation is required. For all other situations the emphasis should be on the use of calculators and computers so as to minimise the drudgery involved and enable students to concentrate more on interpreting the data. Scientific calculators are an obvious tool for the calculation of summary statistics, but a computer will sometimes be more appropriate because of the graphics, editing and more extensive calculating capabilities.

(Band D) Students should learn how to enter and access information in a simple database and appreciate the potential for errors in transcription and coding. They should explore raw data from a variety of sources. (pp. 164, 174, 180)

Database debut

Years 3 to 7

Children build a database containing information about themselves, based on questions which they have designed. Once the data are entered into the computer, they are encouraged to ask the kind of questions that the database can most easily answer, and learn to compare different groupings and so on. This investigation is designed to introduce teachers and children to the use of computers as data storage and manipulation tools.

Features

Teaching and learning

- Personal data
- Students take control of collecting, entering and questioning data
- Using a classroom computer
- Group work

Major mathematical content

- Posing and refining questions
- Collecting and interpreting data
- Opportunity to present data in tables and graphs
- Comparing data using percentages
- Developing measurement concepts
- Finding members of intersections of two sets (those who are in one set AND the other) and unions of two sets (those who are in one set OR the other)

A discussion of the general characteristics of a database and a glossary of terms are in the *Content epilogue.*

Equipment and preparation

- At least one computer continuously available in the classroom and frequent access to a printer
- A simple database creation program with normal *sort* and *search* capabilities
- A graphing option or graphing software is useful
- Experience using a computer database is not essential but it is worth being familiar with the way *records* and *fields* are set up and the types of *sorts* and *searches* in the program you use
- Additional secondhand data for 49 students, with 13 fields of information for each, is available on the software disk under Investigations Data

Expected timing

- The introduction could take three or four sessions, and data entry for your own class a few more
- Once the entries are in, three or four sessions are needed to explore the data, but this could be spread over many months
- If you extend the data collection to the whole school, you have weeks of fascinating work

For more information on database issues, see the introductory sections in the companion *Chance and data* book, *Exploring Real Data*.

A teacher's story

Databases are used widely in the community. I thought it would be educationally valuable for my children to experience some of the typical features of these firsthand. I felt that they would be more likely to understand the whole idea if they first actually created a database of their own filled with information about things of interest to them.

Basically I introduced the features of a computer database in terms of a typical card filing system. I talked about some ways the cards could be *sorted* or *searched* for particular purposes, then suggested they might like to create a database about themselves. The children dictated the pace, so the way the project finally turned out, and is described in this investigation, is the direct result of my own class's ideas, interests and purpose.

You should not expect your children to come up with the same thoughts; the investigation's power is in the personal 'ownership' of the database. However, even though the direction and information may differ, there are likely to be similar opportunities for developing critical thinking, problem-solving skills and an understanding of the considerations and mathematics involved in collecting, manipulating and interpreting data in this way.

Comments

My main concern was that I had no firsthand experience of either using established databases or creating new ones. I need not have been concerned; the children's enthusiasm and initiative soon gave me confidence. In fact, my inexperience turned out to be an advantage. Because I had no preconceived ideas about how the project ought to proceed and only a limited knowledge of the actual program we used, the children seemed to become more directly involved in recognising and solving the various problems which arose and soon began to take control of the project themselves.

In effect they became the 'experts' and I often had to ask them how to carry out a particular procedure. Their growing technical self-reliance allowed me to stand back and observe the development of their understanding and draw attention to particular issues and ideas that I may have missed had I been too engrossed in the practical aspects of gathering and entering data.

Summary of the stages in this investigation

1. **Introducing the idea of a 'database'**
2. **Planning our database**
3. **Setting up our database file**
4. **Entering initial data**
5. **Developing preliminary questions**
6. **Developing questionnaires for the whole school**
7. **Entering data**
8. **Preliminary sorting and searching**
9. **Checking and editing data**
10. **Asking questions: sorting and searching**
11. **Analysing data and reporting results**
12. **Explaining results**
13. **Extensions**

1 Introducing the idea of a 'database'

I introduced the term 'database' as being similar to a card *file** of *records* holding *data* about something; cards which can be sorted or searched. We discussed the kinds of information people might put on cards and for what reasons. I had a range of record cards for children to handle to help the discussion.

> I think that right from the start it's important to focus on the idea of purpose for collecting and storing data.

Who's seen a record file or box of cards with lots of information on each card?

> **My dad's got his customers on cards. He looks to see who didn't pay him.**

> **In the library you can look at the cards to find out which books have got stuff you want.**

Why do people collect the information and put it on the cards?

How do they decide what they want to put on the cards?

Where does the information come from? What do they do with the information?

If all this data is in a file of records in a computer instead of on cards, you can save a lot of time looking for information you want, because you can tell the computer to look through the cards for you.

> I brought in a bunch of the library book records for the children to handle to help them get the idea and I began to use the terms file, record, field and data when talking about them. Any similar set of record cards would do just as well.

*Terms in italics are described in the glossary contained in the *Content epilogue*.

2 Planning our database

The children were asked to think about making a database about themselves on our classroom computer. We discussed the questions we might want to ask and the reasons. Possible uses and a wider audience for the information were also considered.

They chose suitable *field headings* (topics to ask about) and I wrote them on a large sheet of paper. Children copied the headings and individually wrote their own personal data next to the appropriate headings. We ended up with a pile of data sheets from which we created our database.

Why would it be useful to know the type of icecream bar children like?

> **To see what are our favourite ones.**

Who around the school might want to know that?

> **Teachers for sport prizes. The canteen lady could; she could find out what's the best to get to sell.**

> **What about drinks kids like? She'd want to know what drinks to get.**

3 Setting up our database file

I introduced the software. We chose a *filename*, and I entered the *field headings* we had chosen into the database, together with the *type* and *length* of each *field*, based on the class information provided.

You will need to find out how many fields you can have and how much can be stored in each field for the database program you are using, since they do vary. The children need to be given this information so that their choices can be constrained by it. (They easily understood the importance of field size after setting up a field with headings but not enough space for entry of the data.)

The choices of field headings should be left to the children as much as possible. The discussion and decisions about what to include and why, are important parts of the learning process. You ask as many 'why' type questions as you can to focus attention on field choices without dictating.

For my class, I found it wise to guide the students into as few fields as possible, since they wanted to ask everything they could think of. To keep it fairly simple I guided them to use questions to which the limited number of responses could be easily predicted. They tended to want to ask very open questions, and then the data became unwieldy.

I made a strict rule that only the child entering data was allowed to touch the keyboard at any time; too many fingers jabbing at the keys invited frustration and arguments.

How many letters long should we make the field called 'Favourite Sport'?

Mine's tennis but that's only six letters, football would be more.

I've got badminton and that's nine, but what if someone said 'horse riding'? That would need about 12 spaces.

I reckon 15 is best, in case there's one a bit longer.

Could we agree on some short ways to write long things: abbreviations or codes?

This discussion is best managed with around ten children; more than that becomes 'bedlam' when children are actively involved as is necessary for maximum learning value. I used three separate shifts with the rest of the class occupied with an alternative quiet activity on each occasion.

This step is important as part of the children's development of a sense of ownership of the database. They are making the decisions and they are learning how it is set up by discussing and arguing the choices involved. Setting up the database can be a bit tedious but it is essential that teachers do not rush this or override the children and use their superior experience to impose 'better' solutions.

Anticipate problem areas and draw children's attention to these if you can see really major difficulties are likely to result, but allow children to make unwise choices so that they will later find their own problem areas and learn from this. They won't learn as much if you insist on the 'right' decision from the start. For example, I left three blank fields at the end so that 'afterthoughts' could be catered for.

4 Entering initial data

The first records were entered into the database by the older children, using their own personal data. The children became familiar with the program and initial problems and oversights were tackled and solved.

Entering the data is best done in a learning chain. In this way all effectively learn the technique with little teacher involvement.

- The first child enters data with the help of the teacher, while a second child watches.
- The second child then types in data, while a third watches. The first child acts as tutor and gives advice.
- The first child leaves, the third child begins entering data, the fourth child joins the group to watch and the second child takes over the tutor role, and so on.

Different students handled this process more easily than others. Existing social groupings and lack of empathy between students hindered this, so I had to decide on the order using a random draw.

Final form of the record we used

RECORD NUMBER 45

1	FIRST NAME:	MICHELLE
2	SURNAME:	G
3	DATE OF BIRTH:	11:19:81
4	SEX:	F
5	HEIGHT:	132
6	MASS	27.0
7	PETS:	1 CAT
8	EATING PLACE:	CHINESE RESTRNT
9	TWO COLOURS:	PINK BLACK
10	TRAVEL TO S	CAR
11	ROOM NUMBER:	01
12	DRINK:	ICED COFFEE
13	ICECREAM:	DRUMSTICK
14	HOUSE NAME:	S
15	SPORT:	CRICKET
16	LUNCH:	HOT CHIPS
17	YEAR LEVEL:	4

The problem is that we haven't got enough space for everything that might be interesting.

We should have put our 'House name'; it might be good for sports day to have all the house names?

Could we put it instead of one of the other fields?

We have to replace one of the fields to fit 'House name' in. Maybe we could vote on it.

What about instead of 'Lunches'?

No, the canteen needs that!

It's important for children to be encouraged to think about the way the database is going at all times. This can be encouraged by following up children's comments.

5 Developing preliminary questions

My class made out a preliminary set of questions to collect the data from each other. During the collecting of data, the schedule was modified in anticipation of developing a questionnaire for use in other classes. During this step children became aware of the need to ask clear questions. When they supplied their own data, they knew what was expected because they set up the field headings. Asking someone else to supply the information revealed the ambiguities.

Karen didn't know what I meant when I asked, 'What's your date of birth?' But she understood when we asked, 'When's your birthday?'

When I ask, 'What's your favourite icecream?' they think I mean flavour and I have to tell them it's the icecream bar name that we want to know.

6 Developing questionnaires for the whole school

My students refined the ambiguous questions and made a questionnaire which was reproduced for the rest of the school to complete. They gave it to the rest of the school one class at a time, then examined and entered the data. Problems still arose, and were fixed.

Someone in Room 4 wrote 'in my bedroom' for 'favourite eating place'.

What did we mean when we asked the question 'Favourite place to eat out?'

Maybe we could try 'What's your favourite restaurant?'

It's important that you try to avoid giving answers to these problems yourself. Even if the children's efforts seem not much of an improvement, it's better if they actually find out by handing the new questionnaire to another class and then examining the data themselves.

Only reproduce enough questionnaires for one class at a time so that the children can examine the data from each class and refine their questions further if needed.

This step allows children to really clarify their questioning, as writing a question for an unseen audience is different from asking a question face to face. The main purpose of this process is to develop an understanding of the difficulties involved in collecting valid data using questionnaires which purport to find out opinions and preferences.

7 Entering data

Students continued to enter the data as the questionnaires were handed in. Sheets are crossed to show completion. Remember to backup your database often, a good habit for children to learn early.

All the Year 7s have put things like 1.37 centimetres for height. Surely that means 1.37 metres.

That's 1 metre and 37 centimetres or 137 centimetres, right?

Hey, just take out the dot, it gives you the answer! You don't have to add it.

Entering data is time consuming. I rostered the children throughout the day so that the computer was continually in use even during other lessons, but I made sure that individual children would not miss lessons which may have been important for them.

Having children in pairs to enter data gives them the opportunity to discuss the interpretation of data. I found that many were happy to continue the job during lunch and recess breaks.

Listening in to children's discussions while they're interpreting and entering data can give an insight into a whole range of developing skills which might not be directly related to chance and data alone.

Some children timed each other to see how quickly they could enter data.

8 Preliminary sorting and searching

We started to *sort*, *search* and print out after only one or two classes had been entered to discover patterns and unearth possible problem areas.

Practising *searching* allowed children to see that coded entries such as for pets had to be carefully thought out in advance. It was useful to try out searches after about 20 or 30 records were entered so problems were exposed before too much work had been done.

When field lengths are restricted, coded data is useful to maximise the information which can be included in a field. However, trial searches can sometimes expose coding problems which may not have been anticipated in the beginning. Children should be encouraged to solve these themselves.

We used CA for cats, and CAN for canaries, but when we searched for cats, it listed canaries as well.

We should have used CAT. It needs three letters for these ones.

Let's look through the code list and see if we can spot any others which might cause problems.

I found that sorting a field showed up spelling errors and helped the children to fix spelling when entering the rest of the data.

These processes develop problem-solving skills and it is important that you involve the children in this individually and in groups as often as possible, stepping back and allowing them to consider possible solutions whenever possible. They will exercise these skills when they can see that their decisions and opinions are taken seriously and acted upon, and not continuously overridden by the teacher's authority.

I find that the hardest thing to do is to step back and let the children themselves struggle and argue about possible solutions to problems. I have to bite my tongue at times and force myself not to always be the ultimate authority. When I do manage this though, I am invariably surprised at the way children are able to take charge of their learning and make sensible decisions.

9 Checking and editing data

They *searched* the data by class and compared them with class lists to find omissions and errors. Children needed help to organise how to *sort* into alphabetical order to match the order of the class lists so that checking entries was easier. 'Vinler cane' was easy to correct when it was next to 'vanilla cone' in the alphabetical listing.

If your class has access to more than one computer, two or more copies of the finished database could be made for simultaneous sorting and searching on all computers. Make sure you keep a backup master copy which can be kept separate from the working copies. Don't forget to back up regularly throughout.

Entering and interpreting data and correcting the inevitable errors allow children to see that computers are only tools, as accurate as the human operator, not infallible instruments.

10 Asking questions: sorting and searching

Now was the chance to pose questions of the database and to plan how to answer these. The ability to ask good questions evolved over time.

> Young children are fascinated to see personal data on screen or printed out. The sophistication of the questions will vary, but with experience even quite young children can formulate and modify their questions for particular purposes.
>
> I found that children needed most help from me at this point as some database programs are not as easy to ask questions of as others. Planning sorts and searches is quite tricky. We learned together, and I allowed the children to be involved in my own learning process.

What does Jason like to eat for lunch?

How many other children like hot dogs for lunch?

Do more boys than girls (or younger or older children) like hot dogs?

Let's sort lunches into alphabetical order and report the most popular lunches to the canteen.

I wonder if young children like different lunches to older children.

I typed in the searches while the children told me what they wanted to find out.

I want to know how many people liked Coke best. What should we ask the computer to do?

Look for 'equals coke'. Some kids spelt it 'cocke'.

We could ask the computer to search for 'equals Coke' or 'equals Cocke', and it would find both spellings, or we could just ask it to look for 'co' without the 'equals' and that would find all the words with 'co' in them. Would that work do you think?

Wouldn't there be other drinks with 'co' in it? Let's try it and see.

Look, there's 'cordial' too, that starts with 'co'.

What about 'ke'?

I wonder if boys or girls like Coke more. I bet boys do and I bet girls like Fanta better.

Well, what should we ask the computer so we can find that out?

Once children got the general idea of the possibilities of *searches*, I tried to get them to suggest different ideas and to look for easier ways of *searching* and finding what they wanted to know.

> Although having a clear purpose for questioning the database is useful, plain curiosity is also valuable. It can lead to more important hypotheses which can then be tested by formulating more questions.

11 Analysing data and reporting results

Once children understood the type of information the computer could provide, groups or individuals took on 'projects'. They obtained information by searches and sorts, analysed it and reported the results.

I tried to get children to think of audiences for their reports as these determined the types of questions to be answered. All of these were real contexts where the results were hopefully to be acted upon by others.

- The canteen and the health teacher — children's food preferences.
- The sports teacher — the favourite sporting activities of 'houses', genders and age groups.
- The principal — safety: travel to school, the school crossing, parent parking and bike racks.

It is worthwhile using a complete sort and printout. Children can count up frequencies which can then be entered into a graphing program.

> If the program you are using does not have a table or graphing facility, printing out a sorted field enables children to count frequencies and construct their own graphs.
>
> The advantage of this 'long way' is that spelling errors can be spotted. For programs which graph direct from the data, spelling consistency becomes essential. In our data, drumstick was also spelt 'drum stick', 'drwmstick' and 'drumbstick' but the children were able to count all these as drumsticks.
>
> Children's understanding that purpose is important is confirmed when the data is actually used for real purposes. The children were delighted when the canteen altered its order of icecream as a result of the survey.

12 Explaining results

I found that a few more girls than boys get driven to school by car, but lots more boys ride their bikes. It's the other way round for walking.

I asked the children to attempt to explain the results of their investigations.

Some children might like to investigate their results further as a separate project.

Why do you think that might happen?

I should have to ask why they walk or ride.

Could we make another database just for that?

The main idea here is to develop the understanding that the data itself can only show that a relationship appears to exist; it doesn't show why it exists. Children at this age need help from the teacher to see that one result may have a number of explanations, and further questions may clarify results.

More girls than boys chose healthier drinks like milk, fruit juice and water, rather than Coke and Fanta. Some girls immediately said that it shows girls are smarter than boys, but of course it could have meant several other things. Here are some, *not* invented by the girls.

- girls gave what they saw as the more acceptable responses;
- girls are given more direct health advice from their parents;
- girls are more conscious of health affecting their looks;
- girls take more notice during health lessons;
- it could have just happened by chance.

13 Extensions

A: Personal projects based on initial data

Children followed up the questions and hypotheses above as individual, small group or whole class projects.

I was able to make a copy of the original database, keep just the fields and data students wanted such as name, gender, year level, etc., creating new fields for information arising out of the new survey.

How would we go about investigating the reason why girls and boys differ in the way they travel to school?

Could we send a questionnaire home for parents to fill in?

Why don't we make a questionnaire for the kids too, and ask why they travel as they do.

Does it make a difference if it's winter or summer?

What questions could we ask to find out things like that?

Sometimes the children's ideas were too complex to be practically investigated. I used my experience to restrict or redirect when I could see that children might become disappointed if allowed full reign. However, I did make a real effort to try to stand back a little and permit them time and space to discover and solve their own problems.

The purpose of this stage is mainly to allow children to consolidate and develop their understanding of data collection and analysis by following up their personal questions. It may also allow you to assess children's understanding in terms of the statements addressed. Not all children will be ready for or will show interest in these individual investigations.

B: Revising the database

Later in the year, the students decided to update and revise the database, and to make comparisons. We could all see how children's preferences and personal data change over time and children discussed the implications of this.

Do you think it would have made a difference to the 'Favourite sport' question if it was asked in summer or winter?

Have a look at the number of children choosing football each time and see if it changes.

C: Comparing our school with other schools

I found a colleague in another school who was prepared to use our questionnaire in her class, and provide the data on disk. This data (for 49 students, 13 fields in each) is available on the software disk under Investigations Data. The children were keen to see how their data compared to those of children in other areas.

This provided a new dimension for the questionnaire development. There was no way to confirm the data from another school as had been possible for data from the children within our school, so more thought about ambiguity of questions was needed.

Should we change any of the questions? Once we get the data back we won't be able to go next door and ask what someone meant by a question.

I think we should limit the choice to bike, car or walk for the way to school. This makes the question easier for children to answer.

D: Creating individual databases

Once children knew and understood the use of the database program I gave them the freedom to use it to create a database of their own.

I want to make a database about the Solar System.

Where will you get the information you need?

I've got some books from the library that have got everything about it. The number of moons of Uranus is quite different in these books!

When were the books published? Some moons are recent discoveries.

Books go out of date fast, don't they!

E: Other subjects and established databases

At this point you may also wish to investigate the established databases available for the computer you are using. They are a valuable support for study in many subject areas.

> I feel it is important for children's developing understanding that they be able to critically examine an existing database and give opinions about its adequacy as well as to check up on the accuracy of its information from other reference sources.

Content epilogue: background information for teachers

Management of large amounts of personal information in computer databases is becoming such a controversial social issue. Knowing something about the power and limitations of this process seemed an important part of being an informed citizen. Here are most of the special words used in this growing area with some explanation, written for you, not the children.

A glossary of terms related to databases

Comparing results When comparisons involving relationships such as gender or age are included, percentages are usually required to enable sensible comparisons. For older children this may not cause difficulty, but for younger children the concept needs to be developed. I find beginning with 'out of ten' useful. You can then extend the idea to 'out of 100'; 'how many would it be if the proportion liking it stayed the same?'. I've found most children over about 8 years old are able to grasp the idea of proportions broadly enough to make sense of percentages used in a graph or table.

Data The data (plural) are all the bits of information your database holds. So the contents of the **fields** are data, the **field headings** are not.

Field A separate field is assigned for each of the categories of information held on the **records**. These can be best determined by asking the overall question: 'What do we want to know about the "objects" in our **file**?' This then results in specific questions such as, 'Are the children boys or girls?', 'How big are the mammals?' or 'How far are the planets from the sun?' which are then translated into suitable **field headings**.

Field headings When you create a **file**, the **fields** have to be given a title or heading which appears on each **record** and are essentially a translation of the specific questions which resulted from asking 'What do we want to know about the objects in our file?'

Field length You may also be asked to define the length of each **field** when you create the database. This refers to the maximum number of characters (letters, digits and *spaces*) you will need for each **field**. Most programs will not allow this to be increased after the database has been created.

Field type The field type refers to the characteristics of the **data** you wish to enter in a particular field, such as whether it will be words or numbers. Programs may vary a great deal in the way this information is entered, but usually you are given a choice of **field type** when creating the database. The basic three types are *text* (which means the **data** is in words, sentences or numbers that are not measurements or counts of things), *numeric* (numbers) or *date* (which allows you to enter a date in a standard format).

File The file describes the whole set of **records** saved under a single **filename**. For example, 'Bestschool' is the filename the children gave to our database.

Record Each record holds all of the information stored about the individual 'objects' in the database **file**. Our **file** 'Bestschool' holds **records** about all the children in our school, and each **record** is about one particular child.

Searches A search will find all the records which satisfy particular criteria and is the most difficult procedure to master when using databases. You are likely to be the one who needs to master the search procedures before the children can tackle this. Some programs' search procedures may be too complex for children under about 12 years old to manage.

Searches may be limited to one field or apply to more than one field.

In one field, such as 'sports', there may be:

one choice:	'Find all the children who play cricket',
more than one choice, linked by AND:	'Find all the children who play cricket AND T-ball'
more than one choice, linked by OR:	'Find all the children who play cricket OR T-ball'

In two fields, such as 'gender' and 'sports', there may be:

one choice from each field, linked by AND 'Find all those who are boys AND who play cricket'

In the field called gender you ask for boys, and in the field called sports you ask for cricket, and you link them with AND. Your instruction might look like *Gender = "M" AND Sport = "cricket"*, depending on the program and the way it expects searches to be entered. (See diagram opposite.)

one choice from each field, linked by OR 'Find all those who are boys OR who play cricket'

Using OR would find all the records in which 'M' is in the gender field as well as all the records in which 'cricket' is in the sports field, so you would be given the records of all the boys, whatever their sporting preferences, as well as the records of any girls who preferred cricket.

AND narrows a search and may give you *less* than the total number of cases in one of the fields but never more, while OR broadens a search and may give you *more* than the total number of cases in one of the fields but never less. (For more on AND and OR searches, see below.)

Some search commands, often called 'fuzzy' searches, may also look for cases of a particular string of characters anywhere in a field, rather than at the beginning or matching with the entire field.

You can also usually use 'greater than' or 'less than' commands (or operators) to look for records which fill certain criteria to answer particular questions. For example, you may wish to find out if taller children run faster, using data about height and sprint time. You could then search for records of children 'greater than' a certain height and find the average or range of sprint times and compare this to records of children 'less than' a certain height.

Great opportunity for problem solving is provided by the almost unlimited possibilities presented by the search commands in a good database program. It's worth the time and effort involved in coming to grips with this powerful database function.

Sorting The data is sorted by selecting a field, and then the program will sort the records alphabetically or numerically (according to the field type). You can then look through the records on the screen or get a printout of the sorted records as a table of the fields you choose.

A **nested sort** is when the computer sorts on two or more fields. Usually the first field chosen is sorted, then this order is maintained while sorting the records according to the next field chosen. This is just the way words are sorted into order: by the first letter alphabetically, then within those, by second letter, and so on.

Here is an example of a nested sort. Suppose you might want to get a listing of all the ways boys and girls travel to school, and you want the ways to be in alphabetical order for each. You sort first the field for 'gender' so that it sorts into male and female. Then you sort by 'the way of travel' so that within each gender group the records would be sorted into ways of travel. The order of the printout would then be: females by bike, females by car, females who walk; then the males: by bike, by car, then walk.

Validity The idea of 'telling the truth' relates to the validity of questionnaire data. If children's preferences on one occasion are different a few months later, what does it mean? Were they telling the truth on each occasion or did their preference really change? The knowledge that these are unanswered questions which may affect the validity of 'opinion poll' results is an important understanding. Many adults do not realise that the results of published polls are often based on a number of assumptions related to this aspect of validity.

AND, OR and sets

The use of the English words AND and OR in this context is rather special, and corresponds to two ideas of how sets of things can combine.

When we say we want those who play cricket AND T-ball, we mean those who play both.

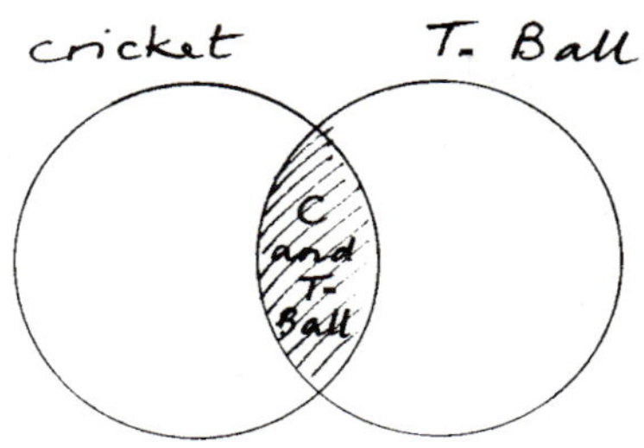

The small shaded area where the circles overlap shows those who belong to both sets; it is given the special name of the 'intersection' of the two sets.

When we say we want those who play cricket OR T-ball, we include those who play cricket, those who play T-ball, and we include those who play both.

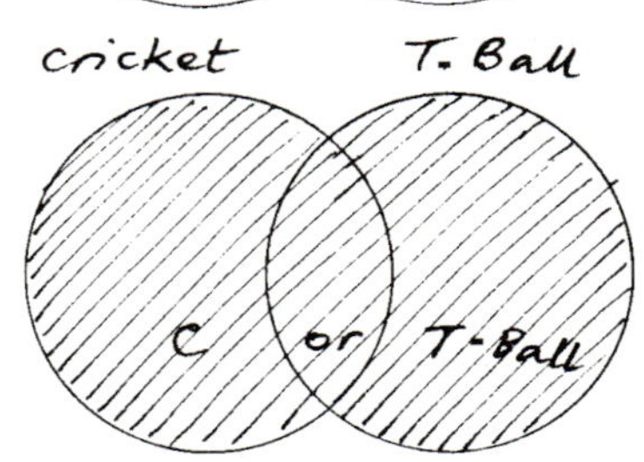

The shaded area that includes both circles shows those who belong to one set or the other, or both; it is given the special name of the 'union' of the two sets.

The English language is quite tricky in this area, and we often seem to say one when we mean the other. For example, to ask for 'the boys AND the girls' is not to ask for people who are both boys and girls! Instead it is to ask for people who are either boys OR girls. The precise mathematical use of the language is only consistent if we talk about characteristics of the people concerned.

Another investigation that deals with this is *Group and sorting rules*.

Pocket money

Years 6 to 8

An initial small scale investigation raises a number of questions about the amount of pocket money received by children in the class. On closer study of the data, children are led to ask more questions: How do you get your money? Do you save some of it? How do you spend it? The investigation is designed to develop an understanding of data gathering as a means of obtaining useful information, and using a database to organise large amounts of data.

Features

Teaching and learning

- Personal data
- Group work
- Use of a computer database
- Questionnaire design, wording, etc.
- Issues of privacy and anonymity
- Handling incomplete and personal data
- Children own the process and are involved in many problem-solving situations
- A story-shell based around a newspaper article and a report to parents

Major mathematical content

- Collecting and interpreting data
- Choosing criteria for collecting data
- Selecting ways of quantifying data
- Making comparisons between data
- Using histograms to compare data sets
- Understanding the idea of the mean
- Reporting numerical results of survey data
- Using percentage to compare proportions

Mathematical content is further discussed in the *Content epilogue*.

Previous experience assumed

- Children should understand percentage as an expression of a proportion and have collected and graphed data on numerous occasions
- They should be familiar with calculating the mean
- They should have already created and used computer databases; if not, include aspects of the introductory *Database debut*

Extra data

- The software disk contains additional data from another class as a source of secondhand data

Expected timing

- About 2 to 3 weeks

A teacher's story

This sequence began as a small exercise in which my Year 6/7 split-class were asked to individually collect, organise and report on data following an assertion that young people receive too much pocket money. (This initial activity was essentially an assessment exercise to enable me to discover the children's level of competence in data collection and analysis. It was purposely non-prescriptive so that the types of choices children made in this context could also be assessed.)

The children's responses and queries which resulted from their data collection led to a suggestion that a full-scale questionnaire should be developed and presented to the whole school so that an informed report could be made to parents as to the pocket money received by different age groups. Children had already been exposed to the use of computer database programs and it was suggested by them at the outset that the computer should be used to process the data they obtained. I found the children were very keen during this sequence and appeared to take a real interest in obtaining and processing data which concerned them so directly, and which had the potential for influencing their parents via the report.

Comments

Once the children had the power of databases at their disposal there was practically no limit to their curiosity. They wanted to ask many questions, and issues of privacy became very important, as they are in general society.

Children realised the extensive amount of interpretation needed to fit complex data into a database. They could see that it was not a 'neat' process, even though the questions were thought out carefully with the database in mind. Thus it is hoped that children will develop a more critical approach to published sources of data.

Some children felt their pocket money might be increased as a result of this survey, others were concerned theirs might be cut! Many girls were quite anxious to show their parents that boys are generally given more pocket money than girls, believing this to advantage their own arguing position for higher pocket money.

Children could see that the outcome might have different consequences for different children. It raised their awareness of the possible variation in purpose for surveys and how different groups may or may not wish to accept a survey as accurately reflecting reality.

Summary of the stages in this investigation

1 Small scale survey

2 Discussing survey results

3 Developing a format for a larger survey

4 Designing and trialling a suitable questionnaire

5 Administering the questionnaire: issues of privacy

6 Designing the database

7 Entering data: handling problems that arise

8 Summarising data and comparing results

9 Using frequency graphs to show comparisons

10 Presenting the findings to a wider audience

11 Extensions

1 Small scale survey

I read this newspaper article to the students and waited for the reaction.

Too much money

Young people of today get too much pocket money. A lot of the problems we have with the youth of today are caused by the amount of pocket money they receive. It makes them think that money is easy to come by and that they should have anything they want. Parents need to take more responsibility and cut down on the amount of pocket money they hand out so that children will learn to appreciate the value of money.

I don't get much, only $2 a week.

I get more than that but I have to pay for my lunch on Monday and Friday.

Let's conduct a survey in our class to find out how much pocket money we each get and what that means. Then we can write an answer to the article.

I suggested they might embarrass someone who maybe doesn't get any pocket money. How could they find out the same information without identifying the people who give the replies?

Let's have a secret ballot... Where you don't have to put your name.

We'll have to write a question on a sheet, fold the answer papers and put them into a box.

2 Discussing survey results

Children asked only two questions for the survey:

- How much pocket money do you get each week?
- Boy or girl?

The anonymous replies were dropped into a box. A small group of students read out the numbers. The results varied between zero and $30, and I wrote them on the board under the headings of 'Boys' and 'Girls'. I then asked them to graph the results.

Often in the past I have left a data gathering exercise at the stage when the simple questions have been asked and the data are in and graphed. I now realise how much potential for learning I have missed! This time I took it much further and a very rich experience followed.

Most made a simple graph showing the amounts received and how many received each amount. A few grouped their data and one child separated girl and boys on the graph. Graphs and reports were displayed on a noticeboard.

The next day a class discussion took place.

Boys get more pocket money than girls — look at Simon's graph.

How could that be explained do you think?

Boys have to pay for more things.

That's not fair — girls need just as much.

Some girls get lots — look one person gets $30 a week.

But I have to pay for everything with that — all my clothes and school stuff and everything.

After a little more discussion, they decided to ask more questions, in order to make a good reply to the newspaper article. They chose to ask questions like 'What do you spend it on?' and 'Do you save some?'

3 Developing a format for a larger survey

How could we find out more information to try to answer the article more fairly?

The students addressed the problems and questions previously discussed in order to design a questionnaire for other classes.

We decided to obtain a lot of data to look at and analyse using a database. We would type up a report which could be sent home to all parents.

We could ask the whole school, not just our class.

Why not make a database for it?

We need to decide what information we're going to collect. And what we're going to do with it.

We can ask what people have to buy with their pocket money as well as how much they get.

Could we tell parents what we find out?

If you are unfamiliar with databases, look at the *Content epilogue* or the investigation *Database debut*. The children were keen to think of questions which might be asked once the initial idea was accepted and they understood that by using the database we could ask up to 20 questions (one for each field). The computer would do a lot of the more difficult calculations and organisation for graphing.

4 Designing and trialling a suitable questionnaire

After the initial discussion, the ideas were becoming too difficult to pull together. I suggested that they each write two lists:
- of all the ways they get money;
- of all that they do with their pocket money;

so that we could put them together and work out questions to cover all the possibilities. These lists were handed to me without names.

I collated this information in my own time and produced a list to help the final development of the questionnaire. The children decided on the questions, keeping in mind what would fit in the database.

All the ways children spent money were listed including how much they spent last week in each category, and extra space was allowed for any other way of spending which had not been listed. Questions were also asked about how much and how often they saved money, and what they were saving for.

Children chose questions to cover receipt of money through 'regular pocket money', 'asking parents', 'birthdays or Christmas', 'special jobs' or other ways.

Children needed to be reminded to include personal information like age, year level and gender so that important comparisons could be made.

When the questionnaire form was finished (using our word processor) I photocopied enough for our class. These were completed as a trial to see if they made sense. Several minor changes were made before the final version was duplicated for the whole school.

5 Administering the questionnaire: issues of privacy

I asked the other teachers' permission to send the questionnaire into their class at a suitable time. I explained the project we were doing, and assured them that the data would be shared with them all. Children went to each room at pre-arranged times. They told the classes that names were not needed, and their own teachers would check that every sheet was sensibly completed.

Children saw that the information obtained should be treated as confidential. They were able to see that children might be embarrassed to write the truth if they know other children would know who they were.

6 Designing the database

This stage can be expected to vary a great deal depending on the amount of experience the class may have had in designing databases for computers. The design of the questionnaire ideally should take into account the types of data required as well as the needs of the computer database in terms of number of fields available and lengths of fields. If the class had previously designed a database, for example using *Database debut*, the actual creation of the database would be a relatively simple exercise.

Each of the questions which covered the way children received money were given a field to show how much was received *each week*. We hoped that the amounts could be calculated to give that information. For example, if they are given $50 for Christmas, that would work out to about $1 per week.

			INCOME (per wk.)			SPENDING (per wk.)			
NAME	SEX	AGE	GIFTS	JOBS	....	SNACKS	MUSIC	VIDEO	....

Questions which covered the spending of money were broken down into categories because there were too many for the database if each were included separately. Here are some examples of our categories:

'saving money', 'lunches', 'clothes', 'video games', 'toys', 'movies', 'snacks' (including icecreams, lollies, drinks and snacks) and 'entertainment' (including comics, magazines, books and tapes/records).

7 Entering data: handling problems that arise

Entering data was carried out by children who were rostered in pairs throughout the day. Each questionnaire was given a number so that data could be checked for typing errors.

It says here that this person gets $100 per week! That can't be true. What do we do?

This person has not completed all the questions. What do we do?

Interpreting data often created problems as it was clear that some information was quite untrue and some questions were unanswered. A class decision was made on whether or not to include any questionnaires that the person entering the data believed to be incorrect. Children decided to include only the data given and leave blank any fields not completed.

It took 3 days to enter the data from the 300 children in our school.

It was important at this stage to find out if the database program would process this information accurately and ignore the numerical blank fields in calculations or if it would treat them the same as entering zero, causing inaccuracies in the results.

Our database only processed and counted those fields with an actual entry. Zero had to be entered as a '0' and blank fields were not counted at all. Unfortunately, not all database programs do this; some automatically enter 'zero' in blank numerical fields. For some databases you need to enter a dash if you want it to be ignored.

Some of the slightly ambiguous questions caused problems. Each of these problems were dealt with as we went along. For example, children had it in mind that they would be able to work out exactly how much children had to spend each week. In fact this turned out to be impossible because many did not give a suitable time span for such questions as:

- Do you just ask your parents when you need money? Yes/No
- About how much?
- About how often?

In fact, many put 'occasionally' or 'sometimes' instead of a definite time span.

It is essential that as teacher you don't over-control the development of the questionnaire. Children seem to learn most from encountering problems which they created, but which are not insurmountable. So I try to anticipate the really important oversights (like forgetting to include fields for gender and year level) but avoid interfering when the outcome is not likely to be so drastic.

8 Summarising data and comparing results

If your class is inexperienced at searching and sorting using a database, integrate the relevant parts of *Database debut* into your teaching at this stage.

Once the data were in, I asked them to write their own ideas for collecting and presenting results. My children were aware of the possible types of searches and sorts.

Small groups got together and compared their ideas, choosing several bits of information they wanted. Then they planned procedures to carry out on the computer. In cases of complex relationships, I helped in formulating the actual way to use the computer program, although simple *sorts* and *searches* were easily completed by children.

We discussed and compared the wide range of results and the most suitable types of tables or graphs in whole class sessions. While children's ideas were used primarily to generate the data summaries and analyses, I took every opportunity to ask questions which might lead to useful results.

How much do you think it would cost for lunch every day?

It shows on this one that he paid $1.50 a day.

How would that change the 'free spending' money he gets each week?

Could the computer take off the lunches from the pocket money and see what happens to how much they get?

If we had any spare fields we could ask the computer to calculate what you're asking and put the result in another field, but we used up all the fields.

It doesn't matter. I'll get the computer to list the fields for pocket money and lunches and I can work it out for myself.

The approach of *Database debut* worked well for classes without the confidence to use the search facility. Basically, the teacher does the typing and children suggest what should be tried. They all discuss the usefulness of the outcome. This method has the advantage that the teacher can learn at the same time as the students.

Results and summaries obtained included children's mean (average) pocket money at each year level and for boys and girls separately for each year level.

Although the numbers of girls and boys, as well as year levels, were fairly equal in the school, children were made aware that straight numerical comparisons could be deceiving and *percentage* data were often more useful. They were also able to see that different ways of grouping the data altered the picture.

Much of the data were obtained by finding the percentage of children receiving each category of pocket money who satisfied other criteria. For example:

- the percentage of children who received over ten dollars per week who also saved their money (or who spent money on clothes);
- the percentage of children who received no regular pocket money but earned it by working;
- the percentage of children who received regular pocket money and also earned other money.

To illustrate the idea of 'mean' (that is, simple average) I gave a number of children various amounts of play money. I then asked them to put it together and share it out so that each got an equal amount.

To show the limitation of the 'mean' for describing what any one individual might receive I showed that for different ranges of amounts, the mean could be the same. For example, both these groups have the same mean (average) pocket money:

A $4, $5, $5, $5, $6

B $1, $1, $5, $9, $9

9 Using frequency graphs to show comparisons

We were able to compare the means for boys and girls to the frequency data to see how the information provided by each gives a slightly different impression. Grouping of the frequency data gave a more dramatic picture of the inequity than did the mean results.

Frequencies for each separate amount of pocket money received per week were awkward and unevenly spaced when graphed. Children needed help to see that grouping the amounts gives a more sensible and useful result. This led to suggestions of broader grouping for comparison between genders.

For example, more girls than boys in my class received $5 or less. More boys than girls received more than $5. The difference between the means was only small, but the frequencies showed strong results.

We set about trying to see how the girls were being disadvantaged. The differences between boys and girls were more noticeable for older children, but then older children received more money. We analysed Years 1 to 3 results using just two groupings ('$1.50 or less', and 'more than $1.50'). Immediately we saw a relationship between boys and girls which corresponded to the inequity found in the senior results.

Pocket money per week

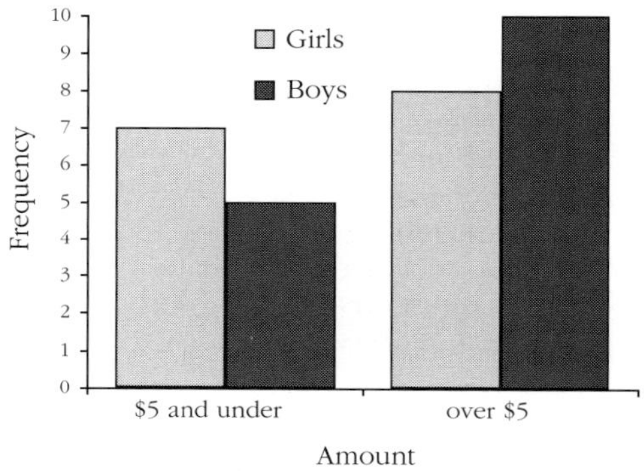

10 Presenting findings to a wider audience

Students discussed and agreed on the aspects which should be 'published' to the rest of the school. There were two aspects of our final results which children thought needed to be reported.

1 There was a great variation in the amounts individual children received at each year level although the mean amount was higher for the higher years. Some children at every year level said they received no pocket money while others said they got more than $10. When all methods of receiving money were combined there was less variation, but still a range.

2 There were big differences in what children said they had to do with the money they received. Some children receiving the lower amounts spent more money on snacks and video games than children who got the higher amounts but saved most of it or spent it on lunches and clothes.

Naturally, your results may be quite different. The children will still need to focus on what they wish to report to the rest of the school community.

This work was carried out in small groups of four to six. Each had parts of the results and discussed what these meant and how they could be reported. A person from each group wrote some information and conclusions and then, as a whole class, we wrote the final report. I used an easel and large sheet of paper so all could see what was being written. I wrote what they wanted but added things if I genuinely thought they were overlooked.

But kids do get lots more now than before. Many of the kids in Year 7 get at least $5 a week, but my mum said she only got about 50 cents.

Do people earn more money now than when your parents were your age?

Things cost a lot less then.

Does that matter when you're thinking about how much children get now compared to 20 years ago?

They considered whether their results were supportive of, or contradictory to, the original claim that young people receive too much pocket money.

The final format of our report, which in fact refuted the argument that children these days received too much pocket money, was typed up and sent home to all parents in the school. Children were then able to discuss at home their own standing in relation to the total school survey.

To further develop the idea of 'interest groups' and reasons why survey results may be distorted, the children were asked if the results could have been interpreted to sound as if we thought children *did* receive too much. What data are included in the report and what are left out can influence greatly the way the report sounds.

Some incomplete results were given, such as mentioning the children who got more than $10 a week as well as birthday presents but not saying that many children got less than one dollar. Others said that we needn't have reported that many of the children who got a lot of money did a lot of jobs, or had to save a lot of their money and were not actually allowed to spend it, or had to buy their lunches and pay for clothes as well.

Most children know of inflation in terms of things going up in price and are also aware from social studies that wages are higher now than 'in the olden days', but often do not associate these two aspects into a concept of 'buying power'. One child suggested that maybe adults just think children get a lot because they compare it with what they got, but forget that things cost a lot more now as well. Another said that maybe not as many children bought things for themselves and their parents gave them things instead.

The idea of children in our parents' day not having as many things can also be looked at in terms of an *inverse* cost difference; watches, radios, record players and TV (when introduced) were all relatively much more expensive and far less available to young people. Therefore, a straight comparison between pocket money then and now, even if general inflation is allowed for, does not give the whole picture.

11 Extensions

Buying power then and now

Some children wished to pursue the idea of comparing 'buying power' with their parents. An investigation similar to this, but incorporating how much things cost when parents were young, could be carried out as an extension to this sequence.

Examining other published surveys

We all became much more aware of magazines and newspaper surveys. They were brought to class and created much discussion and interest as to what we thought must have happened 'behind the scenes' of the published results.

Content epilogue: background information for teachers

Taking a simple question a lot further

This investigation grew out of a rather 'straightforward' survey about the amount of pocket money children got. It reveals the richness that lies just under the surface of many apparently trivial questions. The secret to get at this buried treasure is to probe for variations and differences between groups. Comparing the responses of different groups always leads to interesting results, and to the need to ask more questions. In this investigation we compare gender and younger and older students.

The other technique for getting more from a question is to examine definitions carefully. The concept of 'pocket money' is not as simple as it appears on the surface. Children soon found the money was obtained and spent in a great variety of ways. The idea of 'pocket money' as free spending money was undermined by the need to use it for transport or meals or compulsory savings. These aspects led to the use of the database to sort out the many questions that followed.

Issues of privacy and control of data

The vast amounts of data that presently exist on computer databases about each adult in the community raise many issues about the accuracy of the data and the ways it could be misused. These matters are regularly discussed in social studies classes, but they will inevitably arise here. Questions of how the data will be used, and whether the public (in the case above, the school community) gets access to a fairly complete report, are vital.

Whether a group in the community will recognise the final report as a reasonable summary of the facts will often depend on the point of view they would like it to support. Thus many children would not like to take home a report that suggests that their pocket money is well over the average for their school community; yet many would be invited to do so.

Use of summary statistics and graphs for comparisons

The simplest way to compare something, such as pocket money, for two groups is to calculate the average (mean) for each group. Yet this may hide a vast amount of other information.

The most useful way to show how two groups compare is to use two histograms, like the ones shown for weekly pocket money for boys and girls in one class.

These can be underneath one another, like these shown on the right, or in side-by-side columns, as in the graph below. They can also be displayed in back-to-back horizontal format like the graph on the next page.

Pocket money per week: boys

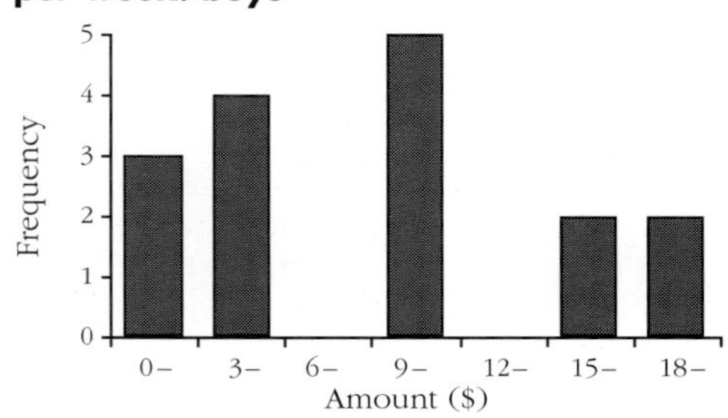

Pocket money per week: girls

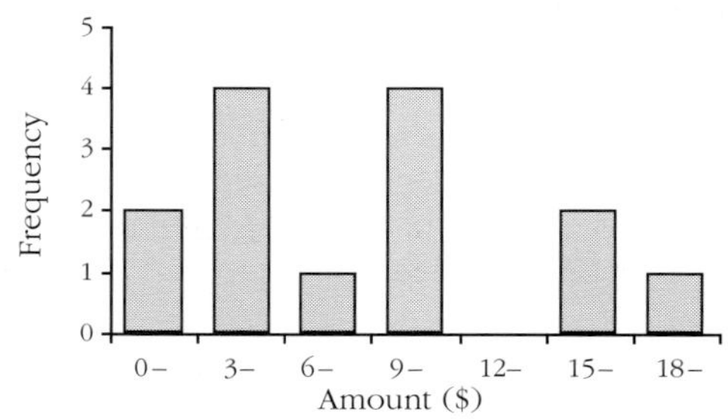

Pocket money per week (side-by-side column graph)

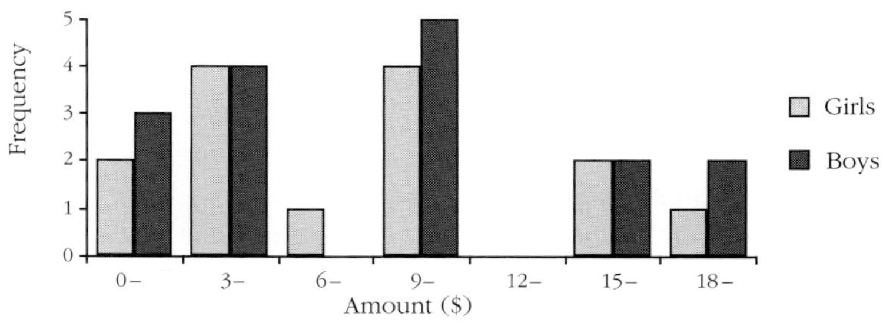

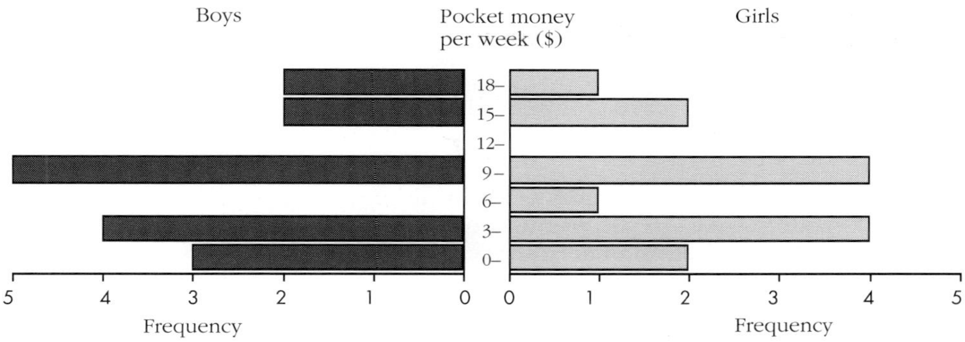

By drawing appropriate graphs, the relationship between the two groups can be more easily seen. One group may be spread out much more than another, or the overlap may be so large as to make comparisons useless. Girls' pocket money in the graphs shown has a slightly narrower spread and a smaller mean. Boys' is skewed to the right, due to a few people with very large amounts of pocket money. As a result of these few, the boys will have a larger mean than the girls. However, if these few are omitted, the mean of boys will be less than the mean of girls.

Another statistic that is less affected by extreme cases is the median: the value in the middle. For the class shown, the median for girls is just below the median for boys.

The use of any single number, the median or the mean, prevents the whole picture from being seen. Only the use of side-by-side frequency graphs (called histograms) can really show the true picture — even though that is simplified when the data is put into groups.

If there are under about 100 items of data, the stem-and-leaf plot works well, since it shows the distributions of the groups' data without losing any actual numbers. For more about stem-and-leaf plots, see *Bikes, monkey bars and skeletons*.

Percentages can mislead

In June 1992 we were told that 48% of youths were unemployed. In fact, it was 48% of a particular age group of those youths registered as actively seeking work who were unemployed. Only 15% of youths in that age group were unemployed. This was tragic, but the headlines tried to make it seem far worse. Of course many may not have been counted because they had given up looking for work, or were in school because there was nothing else to do.

The press is frequently guilty of reporting that a percentage of some group is in trouble for some reason. The group from which the percentage is drawn is often not adequately described. Students need to be on the lookout for such cases and, in reporting their own findings, they need to define clearly the group from which they have taken a percentage.

General data handling issues

The introductory notes to the MCTP *Chance and Data: Exploring Real Data* explore both the general concepts and issues as well as particular statistical techniques for handling the data.

Computer databases

There are many variations available. The most common of these would be MS Works and Excel. A version for MS DOS is FORTE, available by contacting Satchel Software, Angle Park Computing Centre, South Australia.

A recommended Apple IIe version, titled FrEdbase, is available through public domain channels.

Weather patterns

Years 8 to 12

How much do your students know about the weather in the rest of Australia? Certainly they see the whole country on a weather map at the end of TV news. In this exploration of secondary data, students learn much more about their own nation, and also learn how to use statistical methods to make comparisons.

Features

Teaching and learning

- Uses real data
- Links with other subjects (geography, science)
- Estimations
- Computer use
- Group work and presentation

Major mathematical content

- Interpretation of tables and graphs
- Computer simulations
- Probability based on long-term data
- Use of measures of central tendency and spread
- Statistical distributions

Mathematical content is further discussed in the *Content epilogue*.

Equipment and preparation

- The software disk includes:
  - a complete set of daily rainfall and maximum and minimum temperatures for each day in 1990 for State and Territory capitals (essential)
  - a complete set of monthly averages for State and Territory capitals, and 40 other weather stations around Australia. Instead you could use the publication *What's the weather...?* (Bureau of Meteorology) which includes, along with monthly averages, weather records and other interesting information about each capital's weather history
- A database program is an advantage to make use of the information on the disk
- A summary of the daily averages is given in the *Climate almanac* publications available free from the Bureau of Meteorology for the capital cities
- The video described in the investigation *Video data* shows radar recordings of several severe storms and one tropical cyclone
- The NATMAP publication, *Atlas of Australian resources: Third series, Vol. 4: Climate* (about $20 from government bookshops) is excellent as it has large-scale maps and background material

Expected timing

- From two or three periods to several weeks

A teacher's story

Everyone talks about the weather, but no one does anything about it! However, I decided that it was a good vehicle, not only to help students learn more about statistics and probability, but also to increase their knowledge about the climate in different parts of their own country.

I started with Worksheets 1 and 2, where students have to use their general geographic knowledge to identify temperature and rainfall graphs for each of Australia's eight capital cities.

The problem solving and communication of this activity was a highlight, focusing on the main ideas of the unit, while also showing how much they knew about our climate. A matching game, included as a computer extension (IBM version only), increased their ability to interpret graphs and their knowledge of the Australian climate.

Next I led them through a series of investigations using long-term monthly averages of weather data, allowing them to focus on interesting features of the climate in their own capital city, and to compare the climate at different times of the year and in different cities. We used histograms (for rainfall) and line graphs (for temperatures). Along the way I introduced two computer simulations, one for rainy days and one for temperatures, and they began to recognise the binomial and normal distributions.

We used the daily records of weather data from 1990 to compare the fluctuations of daily data with long-term averages. To assist the comparisons I reminded them about box-plots.

As a longer term project I asked groups to work on presenting a detailed study of the climate of a capital city in another State or Territory.

Comments

I learned a lot about Australia's geography and climate, and so did the students. I was really pleased to see the way maths and geography were integrated. I asked the geography teacher to help explain the results we obtained.

The IBM computer game *Weather match* was great fun and helped hold the interest of students.

Summary of the stages in this investigation

1 Climates across Australia

2 Weather match: a computer program

3 Long-term monthly averages in each city

4 Computer simulation of temperature

5 Computer simulation of number of rainy days

6 Ten wet days in a row!

7 Extensions

1 Climates across Australia

Temperature graphs

I handed out Worksheet 1, which is a set of graphs showing the average daily high and low temperature for each month for eight places around Australia. Each graph refers to one of the State and Territory capitals. The places are Adelaide, Brisbane, Canberra, Darwin, Hobart, Melbourne, Perth and Sydney.

The 'matching' activity is a great way to start this topic, and develop interest in both weather and climate around Australia.

Which graph shows which city?

This task works very well in groups. The necessary geographical knowledge is usually available in a group, and you should feel free to answer any questions about which city is further north, and so on.

You might find it useful to ask some interpretive questions about the graphs, such as the following.

Once the groups of four have agreed on an answer, combine groups to share answers, argue out disagreements, and eventually obtain a class consensus. You should not need to tell them the solution since they will almost certainly figure it out for themselves. However the answers are:

1 P   2 D   3 A   4 B   5 S   6 C   7 M   8 H

Which city is the hottest? Which is the coldest? When does this happen? Why?

Which has the greatest range of temperatures? Which has the least? Why?

Which is the most pleasant place to live? Near which cities does it snow? When?

Rainfall graphs

Worksheet 2 is a set of graphs showing the average rainfall for each month of the year in the same state and territory capitals. I again asked the students which is which?

The process is identical to that used with temperature graphs. It will teach students a lot about the climate across Australia. The answers are the same; the two sheets have been designed so they can be placed one over the other as a single summary of the climate in that city. You can demonstrate this with transparencies of the sheets, possibly on the overhead projector. This is also the form of the graphs on the computer program that follows.

2 Weather match: a computer program

The IBM software disk includes a special weather game, designed for up to three students to use at a time.

The disk has the average monthly temperatures (maximum and minimum), rainfall and number of rainy days for each month for 49 different weather stations (around Australia and three offshore islands — see the shaded box below). The program selects four of these, and calls them A, B, C and D. It displays the information in graphical form on the screen. It also shows the locations of the four (using dots) on a small map.

Your task is to match the dots on the map to the graphs (A to D). When you have entered all your guesses, you will get the names of the places.

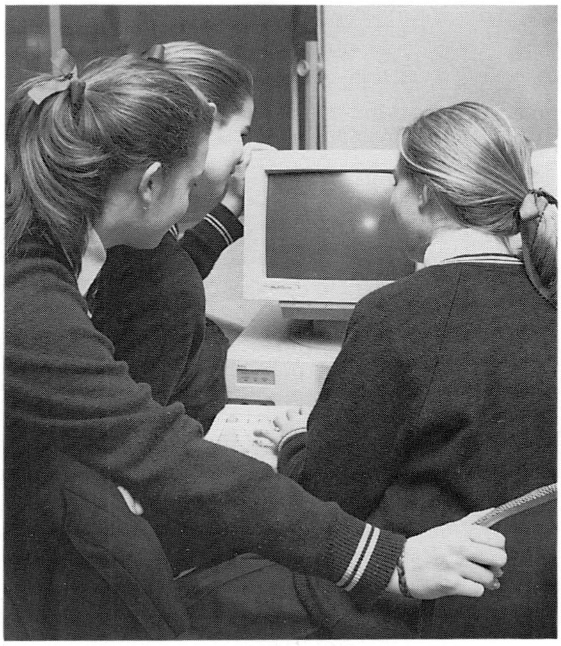

3 Long-term monthly averages in each city

I organised the groups to look at the long-term averages for one capital city.

There is a distinct statistical long-term pattern of weather for each city in Australia. This data is available in several forms.

- The *Investigations data* section on the software disk has monthly averages for each of nine cities (the eight capitals plus Alice Springs) based, in most cases, on about 100 years of records.
- Worksheet 3 provides temperature data for the eight capital cities for each month of the year.
- The computer game (see above) contains ASCII format records of climatic averages for the sites shown in the shaded box.
- The free publication *What's the weather...?*, available from the Bureau of Meteorology, includes monthly averages for State and Territory capitals.

Challenge students using these data sources to build an understanding and an overview of at least one city.

The *Weather match* game has ASCII format records of climatic averages for:

WA	Perth Esperance Forrest Carnarvon Port Hedland Derby	Hall's Creek Kalgoorlie Geraldton Mundiwindi Marble Bar
NT	Darwin Alice Springs	Katherine Daly Waters
SA	Oodnadatta Port Pirie Sterling	Ceduna Adelaide Mt Gambier
Qld	Thursday Is. Townsville Mackay Emerald Charleville	Innisfail Cloncurry Brisbane Normanton
NSW	Wilcannia Griffith Sydney Kempsey	Narrabri Katoomba Kiandra
ACT	Canberra	
Vic.	Melbourne Horsham	Geelong Wilson's Prom
Tas.	Burnie Hobart Nauru Macquarie Is.	Strathgordon Bicheno Norfolk Is.

4 Computer simulation of temperature

Temperature is roughly normally distributed, with some possible skewness. The graph on the software disk (like the one on the right) shows both the normal curve and a five-column histogram based on the 14th, 50th and 86th percentiles.

Why 14 and 86? Values under the 14th percentile can be expected to occur at least once in 7 days, as can values over the 86th percentile. The other reason is that the Bureau of Meteorology has chosen to provide these figures for their usefulness in predicting weekly weather.

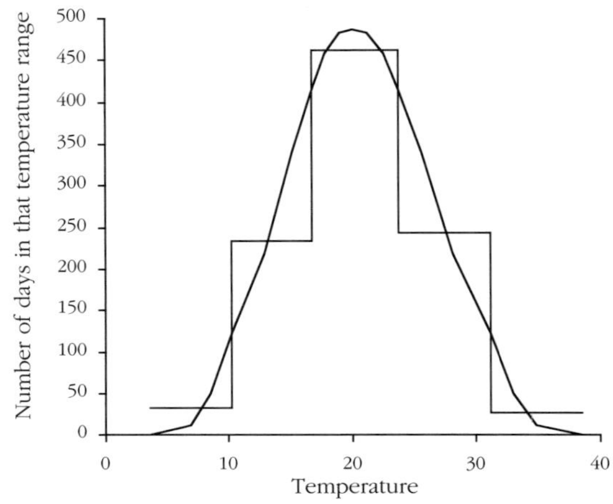

The temperatures for 10 days or 1 month (30 days) can be simulated by selecting from a normal distribution using the computer program, if available.

Demonstration mode

I used the 14th, 50th and 86th percentiles from our own city and current month (from Worksheet 3). We ran just ten trials, representing 10 days.

The program shows the simulated temperatures for 10 days. We compared them with the real temperatures for the last 10 days. It also graphs the distribution of the temperatures in the form of a five-column histogram, and shows a normal curve with the same mean and spread.

We increased the number of trials to 30, to simulate 1 month.

Simulating a real month

On Worksheet 3 are the 14th, 50th and 86th percentiles of the long-term averages of the daily maximum temperatures for each month for each of the eight capital cities. Here is how we used this data.

- Students chose one city and 1 month, and entered those three numbers into the program. They ran the program for 30 days. A histogram shows the distribution of temperatures for a typical month.
- Students compared this distribution with the real data they collected or with the Weather Bureau data.
- They ran it several times to see how much it changed from one run to another.

Substantial variation is possible for any 1 month; the variation is masked when you have only the long-term average and spread. This, of course, is what really happens with the weather!

The shape of the distribution

If students run the program for 100 trials or more, they will get a good idea of the shape of the distribution of temperatures (in the long run). The theoretical shape of the distribution is the normal curve, shown on the screen.

How many trials do you need before the histogram starts to get close to the shape of the normal curve?

5 Computer simulation of number of rainy days

Demonstration mode

I used the *Climate almanac* for our capital city to look up the long-term frequency of rain for this month. We found the average of the values. This we used as the probability that it would rain on any day in that period.

> Alternatives:
> - The long-term averages on the software disk give the long-term average number of rainy days for each month for each of the nine cities. Divide this by the number of days in the month to get the probability of rain on any day in that month.
> - If you have only the book *What's the weather...?*, divide the given average number of rainy days by the number of days in the month to get the probability for that month.

> The probability that it will rain on any day in any month can be theoretically worked out using the average long-term frequency of rain for the month. These values are found in the *Climate almanac* for your capital. To find the average number (or expected number) of rainy days out of 10 days simply multiply by 10. (The formula is $E(x) = pn$.)
>
> Another approach that gives a much more realistic idea of what can actually happen is to use the probability that it will rain (from the tables in the almanac) and simulate the rain-or-not-rain (binomial) situation.

We ran the program once, for one 10-day period. It showed which days were wet and which were dry. We counted the number of each, and entered them into a frequency table.

We ran the program several times. We added the new values to the frequency table and the graph showed the distribution of the number of wet days.

What can we learn about how the number of rainy days changes?

> The number of rainy days will probably have about the expected average or central value, but will have quite a large spread, showing that exceptionally dry and exceptionally wet periods may occur quite often.

We ran the program many times.

How can we describe how the number of rainy days are spread by using the simulation?

> The distribution is binomial, and is only symmetrical for $p = 0.5$. The greater the number of trials the narrower the curve.

6 Ten wet days in a row!

In January 1992 Melbourne experienced a record number of wet days in a row; it sure spoiled the holidays! The 'probability of rain' for any one day in January in Melbourne is about 0.3.

What is the chance that 10 days in a row will be wet, using the simple model above?

Is the weather on one day really independent of the weather the day before? If it is not, then the simple model needs improving.

> Using the multiplication principle, the chance of 10 wet January days in a row is easily found on a calculator:
>
> $0.3 \times 0.3 \times 0.3 \times 0.3 \times 0.3 \times 0.3 \times 0.3 \times 0.3 \times 0.3 \times 0.3 = (0.3)^{10}$
>
> $= 0.000\,005\,904\,9$
>
> The reciprocal of this fraction is 169 350, meaning that the chance of the next 10 days being wet is one in 169 350 January days!

How does the chance of 10 wet days in a row change for different months in different cities?

Students compare $(0.1)^{10}$, $(0.2)^{10}$, $(0.3)^{10}$,..., $(0.9)^{10}$, expressing each as 1 in n.

> Answers: 1 in 10^{10}, 9.8 x 106, 169 350, 9537, 1024, 165, 35, 9, 3

7 Extensions

A: Describe the climate of one state capital: a group project

Each group can investigate one city. Use the data to draw graphs and help describe as clearly as possible the city you are looking at. Your report to the class will need to be clear and easy to understand, and tell us what the climate is like in that city.

This is an extended problem in data interpretation; there is a mass of data and it needs to be organised and described simply and easily. Graphs are the best tool, but they need accompanying descriptions about the whole picture they portray.

The data on the software disk (and the book *What's the weather ...?*) have the information about the climates for all state capitals taken from many years of readings. There are:
- average maximum and minimum temperatures for each month of the year (these were used to draw the graphs on Worksheets 1 and 2);
- the median rainfall for that month;
- the number of days on which rain falls.

B: What happened in 1990?

Looking at one month in our city

On the disk is a set of data that shows, for each state and territory capital and Alice Springs in 1990:
- rainfall for each day
- maximum temperature for each day
- minimum temperature for each day

This data could be printed for analysis or imported into a database program (such as MS Works or Excel or FrEdbase) to assist in computer manipulation of the information.

We compared the averages and ranges of maximum temperature and minimum temperature for the 1990 data.

On how many days in the month was there rain in 1990?

How do the amounts of rain that fell compare?

How are the maximum temperatures distributed?

To show this we put the maximum temperatures in order, smallest to largest, then used the 25th, 50th and 75th percentiles (the quartiles) to draw a box-plot (see the *Content epilogue* and the next section).

Box-plots

I handed out Worksheet 4, which is designed to help students use and interpret box-plots. Average maximum daily temperatures for Perth and Darwin in January are given. The ordered data shows the median, quartiles and range used to draw the box-plots.

This section is included to enable you to discuss the use of box-plots to compare two distributions.

Which has the higher middle value? Which has the greater range?

Which has the greater spread? How do the quartiles show this?

Why are Perth and Darwin so different in January?

Compare different months in 1990 for a city

Since the disk has daily weather data from nine cities from 1990, I split the class into nine groups and organised each group to work with data from a different city. Each group was asked to provide a verbal and written report of the answers to the questions on Worksheet 5, for later presentation to the class. The groups can divide up the task, and use technology wherever possible to speed the process.

This work took the groups about two class periods, with some work at home on the actual writing up of the presentation.

Comparing the actual weather to the 'average' weather

The tasks were to describe the 1990 weather for each month in terms of how it compared to the 'normal' long-term weather pattern for that month. These tasks are described on Worksheet 5.

This task took about one session and the reporting took about half a session.

Presentation of reports and discussion

I allocated one further session to the presentation of reports, and the discussion of the changes over the year in each city and comparisons among the cities.

How did the climate of Darwin compare to that of Perth?

This comparative use of statistics will require a fair amount of care with interpretation. In most cases students should appeal to the data (in the form of graphs, which display both central tendencies and range) rather than to averages.

Content epilogue: background information for teachers

Column graphs and histograms

Histograms are used to graph frequency tables of continuous data; for example, the frequency of temperatures between 0 and 10, 10 and 20, 20 and 30, etc. They should not be confused with column graphs, which are just values of a measurement for different category values; for example, the total rainfall for the month or the temperature on different days of the week.

Histograms are normally drawn as columns of equal width. The horizontal axis should be a number line, and the columns should touch, since every possible value of the continuous variable (such as temperature) must fit into one of the columns.

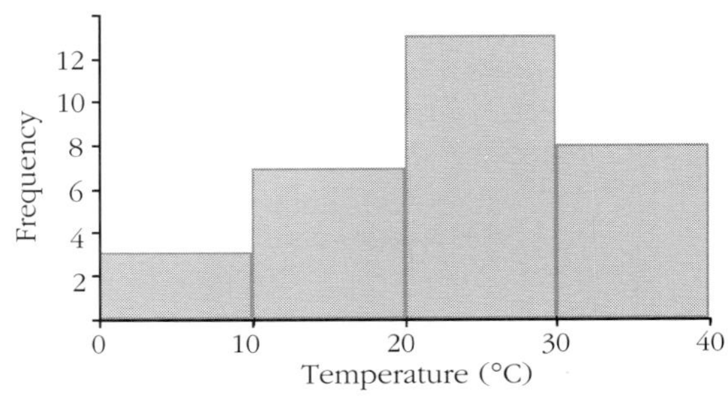

Column graphs represent measures of category values, so they need not have numbers on the horizontal axis, and, if they are numbers, they need not be spaced as in a number line.

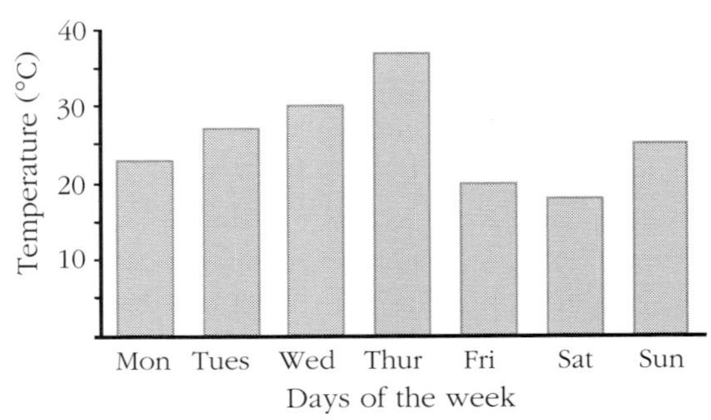

Box-plots

This is a pictorial way of showing a pattern of distribution by using five specially selected numbers: the extremes (lowest and highest), the middle value, and the quarter and three-quarter way numbers. These last three are called the median (middle), and the lower and upper quartiles.

Here are seven numbers arranged in order:

12 13 13 14 16 17 18

The five numbers needed to construct the box-plot are shown. Notice that if there is no number at quarter or three-quarter way you take the average of the numbers each side. A box-plot is drawn using these five numbers.

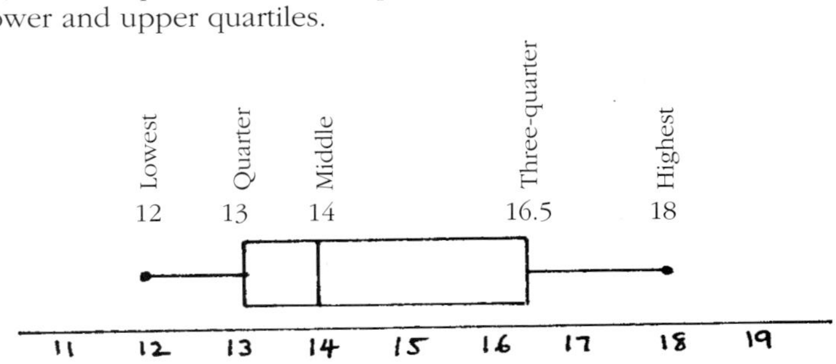

Percentiles

Imagine that all the values in a table are arranged into order from least to greatest, numbered from 1 up, and given percentages according to how far up the list from 0% to 100% they are.

The median (half-way) will normally be at the 50% stage, and it is called the 50th percentile for this reason. The lower quartile is called the 25th percentile, and the upper quartile the 75th percentile. If you were interested in the temperatures over 20 days, then the four coldest ones would be under the 25th percentile, and the four hottest would be over the 75th percentile.

The values on Worksheet 3 are the 14th, 50th and 86th percentiles. These percentiles have been chosen to relate to a week: in the long run, the coldest day in a week (one in seven temperatures) can be expected to be under the 14th percentile, and the hottest day over the 86th percentile.

Using long-term statistics for probability estimates

The Bureau of Meteorology has collected data from many stations for well over 100 years. For most days in those years they have over 100 records of whether it rained or not. (The exception is 29 February.) So we have a lot of information on relative frequency of rain; it is reasonable to use this as an estimate of the probability of rain on that day.

The binomial distribution: rainy days

The simulation of rainy days has two possibilities for each day in the week: wet or dry. The probability of rain is the same each day. These two conditions make this a binomial distribution. Taking the week as a whole, the probabilities for a given number of wet days, from 0 to 7, can be found from this formula, where w and d are the probabilities of a wet and a dry day.

$$(w + d)^7 = w^7 + 7w^6d + 21w^5d^2 + 35w^4d^3 + 35w^3d^4 + 21w^2d^5 + 7wd^6 + d^7$$

For example, if $w = 0.5$, and hence $d = 0.5$, w^7 gives the probability of 7 wet days in a week. This is 0.008 (1 in 128), so in 100 days we expect it to happen $100 \times 0.008 = 0.8$ times (possibly once). In contrast, $35w^4d^3$ gives the probability of four wet and three dry days. This is 0.27, so in 100 weeks we expect that to happen 27 times.

The values are symmetrical when $w = d$. Where this is not the case, the values will be higher on the side of wet (if $w > d$) or dry (if $w < d$). The binomial distribution will be 'skewed' because of this.

The normal distribution: temperatures

The simulation of temperature distribution is based on the assumption that temperatures are normally distributed. This is fairly close to what happens in reality. The simulation is modified to allow for the slightly skewed distributions that often occur.

The normal distribution is usually described by its mean and standard deviation. Because we have information from the Bureau of Meteorology about the 14th, 50th and 86th percentiles (see above), we have designed the program to use this information to determine the shape of the distribution. (If the 14th and 86th are not equally spaced from the 50th, then the distribution will be skewed.)

The range of the normal distribution may reasonably be defined as the extent of three standard deviations each side of the mean as shown in the diagram on the right. The temperatures are chosen at random from within five ranges from the spread of the normal distribution, so defined. After a large number of trials, the shape of this five column histogram should closely agree with the shape of the normal curve. However, if only a few trials (that is, days) are used, they may in no way conform to expectations! This is closely reflected in the real world, and temperatures in the short run are 'all over the place'.

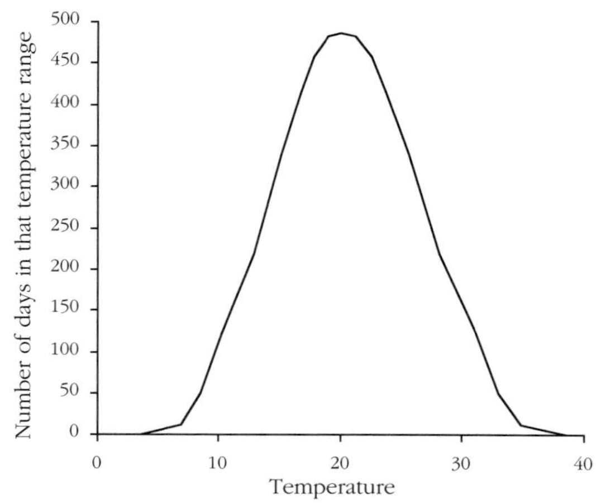

You could test how well the real data from 1990 matches this distribution by taking a number of days during a month, and graphing the real temperatures. Compare the average temperature with the stated long-term monthly average, and the shape of the distribution with the normal distribution.

Worksheet 1: Temperature graphs for eight cities

Can you match these graphs to the cities of Adelaide, Brisbane, Canberra, Darwin, Hobart, Melbourne, Perth, Sydney?

City 1

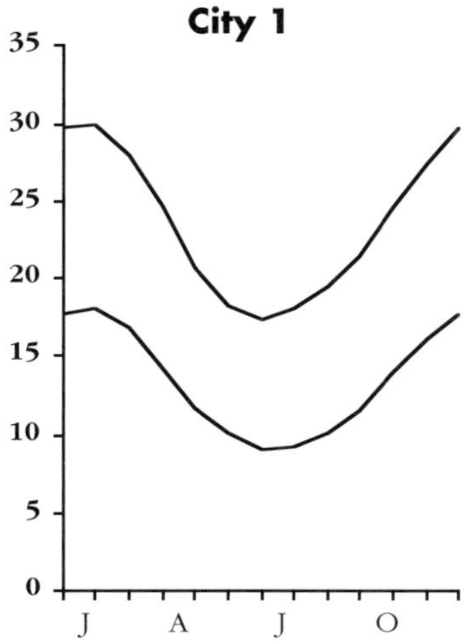

City 2

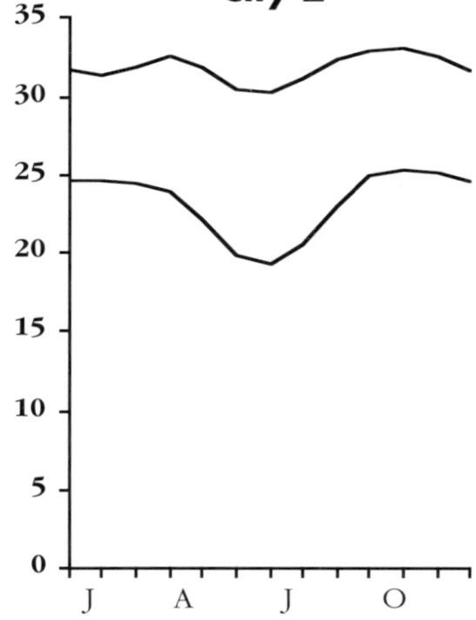

City 3

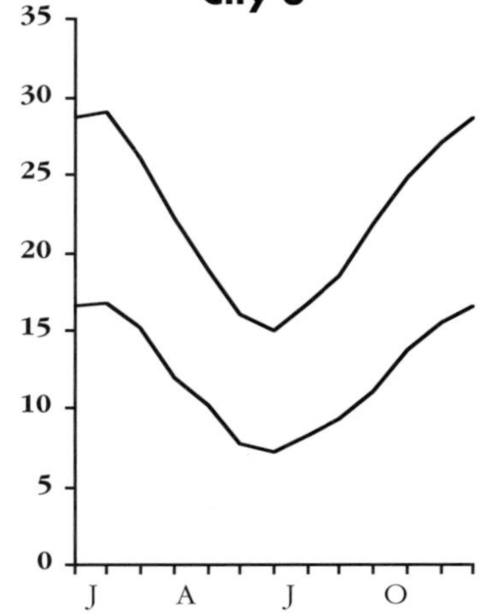

City 4

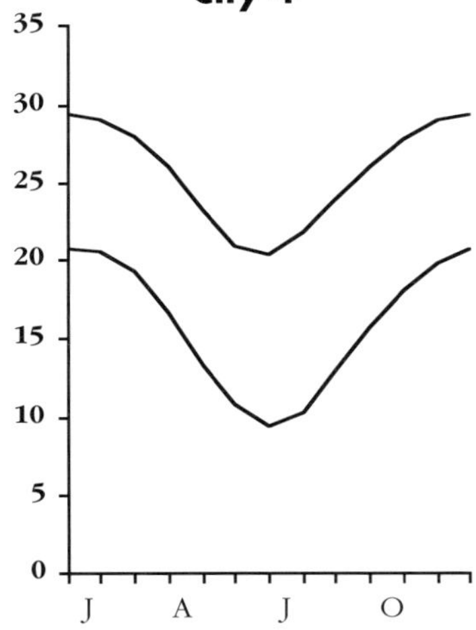

City 5

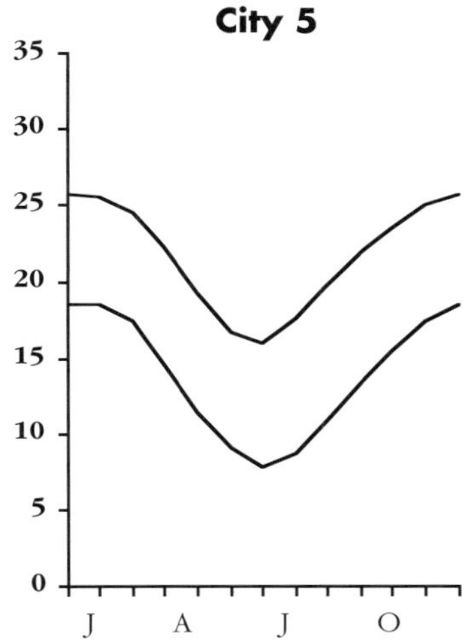

City 6

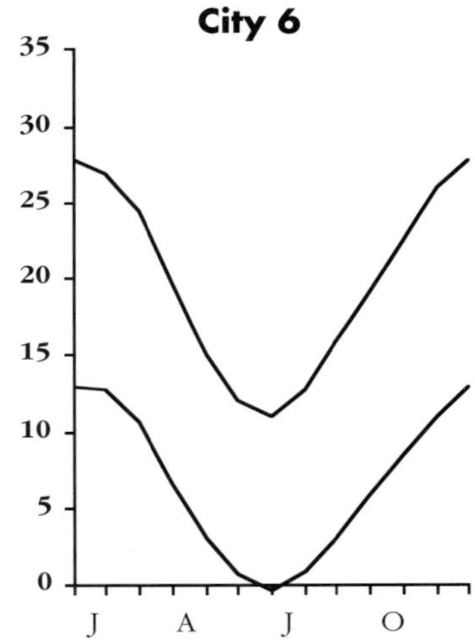

City 7

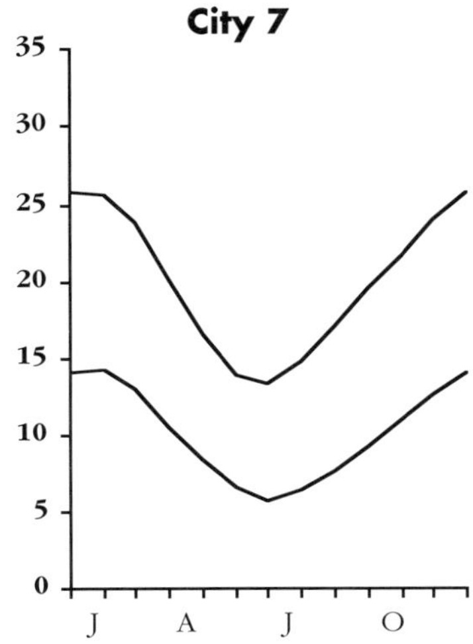

City 8

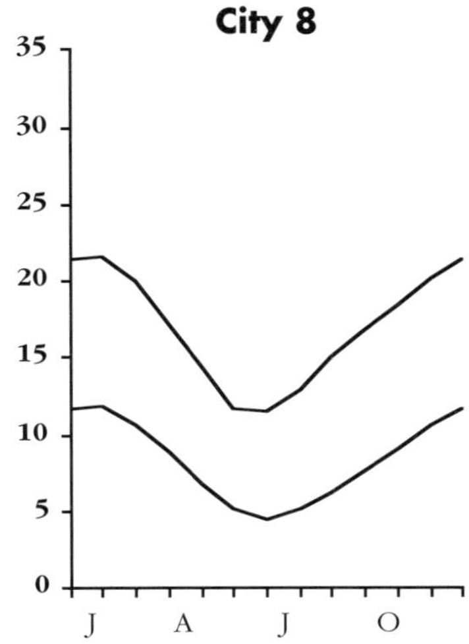

Worksheet 2: Rainfall graphs for eight cities

Can you match these graphs to the cities of Adelaide, Brisbane, Canberra, Darwin, Hobart, Melbourne, Perth, Sydney?

City 1

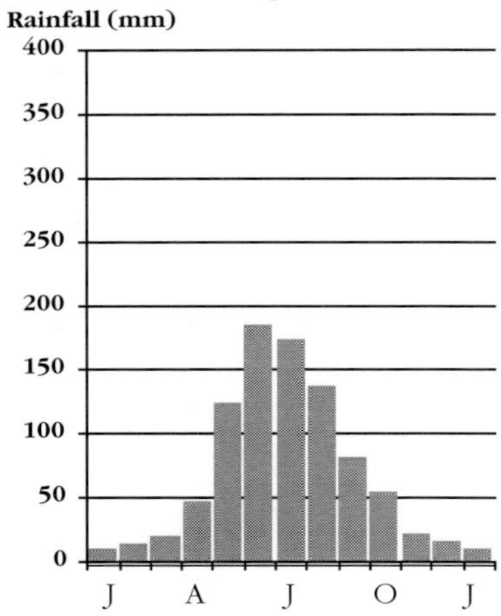

City 2

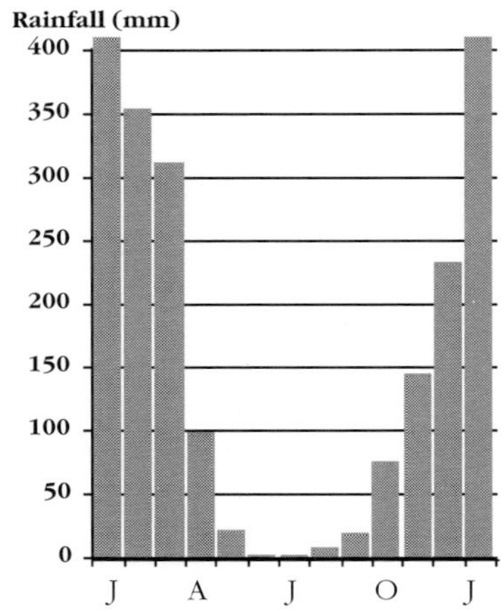

City 3

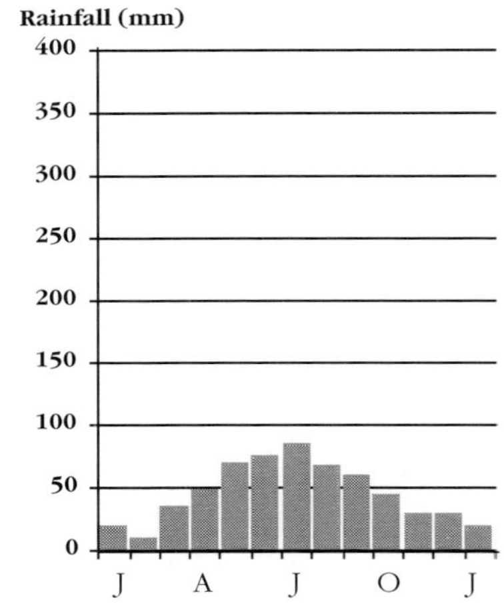

City 4

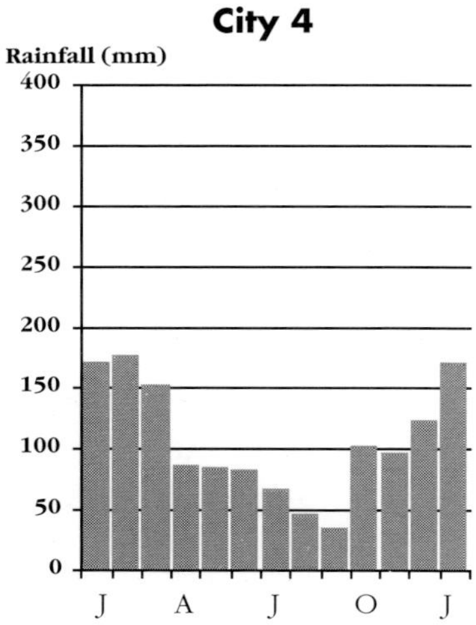

City 5

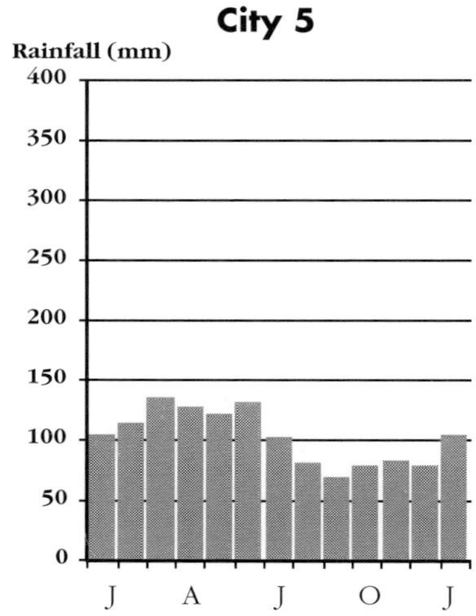

City 6

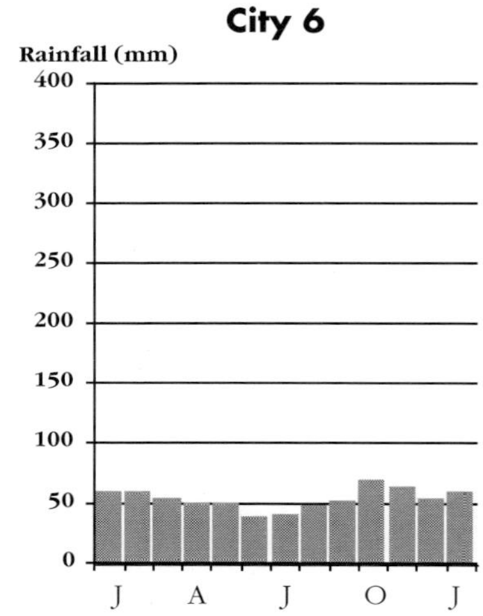

City 7

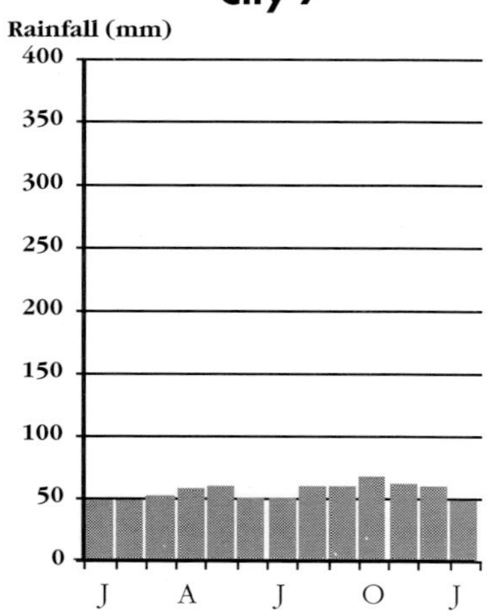

City 8

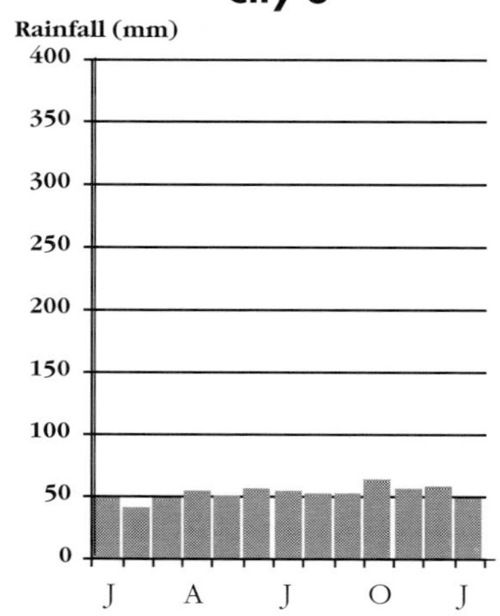

Worksheet 3: Weather patterns

Monthly averages of maximum temperatures (°C) and number of rainy days

	JAN.	FEB.	MAR.	APR.	MAY	JUN.	JUL.	AUG.	SEP.	OCT.	NOV.	DEC.
HOBART												
14th percentile	17.9	17.8	16.5	14.4	2.1	10.0	9.8	10.4	11.7	13.2	14.7	16.2
50th percentile	22.3	22.2	20.6	18.1	15.0	12.8	12.2	13.2	15.1	17.2	18.8	20.3
86th percentile	27.4	26.8	25.1	22.2	18.0	15.6	14.6	16.0	18.0	21.1	23.0	24.6
Rainy days	9	8	10	11	13	11	13	15	14	13	14	13
CANBERRA												
14th percentile	22.2	21.8	20.0	15.8	12.0	9.7	8.8	10.0	12.3	15.0	17.4	20.6
50th percentile	27.8	26.9	24.4	19.7	15.1	12.0	11.1	12.8	15.9	19.1	22.5	26.0
86th percentile	33.4	32.1	28.8	23.5	18.6	14.4	13.3	15.5	19.7	23.9	27.8	31.3
Rainy days	8	7	7	8	9	9	10	11	10	11	10	8
DARWIN												
14th percentile	29.8	29.6	30.2	31.3	30.3	28.7	28.7	29.6	30.8	31.7	31.8	30.9
50th percentile	31.7	31.3	31.8	32.6	31.9	30.5	30.3	31.2	32.4	33.0	33.1	32.5
86th percentile	33.4	32.9	33.4	34.1	33.5	32.2	32.1	32.9	34.2	34.4	34.3	34.0
Rainy days	21	20	19	9	2	0	1	1	2	7	12	16
PERTH												
14th percentile	25.7	26.1	24.7	21.2	18.3	16.3	15.6	15.9	17.2	18.7	21.0	23.3
50th percentile	31.4	31.7	29.5	25.2	21.4	18.8	17.7	18.3	20.0	22.3	25.4	28.5
86th percentile	37.5	37.2	34.7	29.7	24.6	21.2	19.7	20.7	23.1	26.8	30.7	34.6
Rainy days	3	3	4	8	13	18	18	16	13	10	7	4
ADELAIDE												
14th percentile	21.9	22.3	20.8	18.2	15.6	13.8	13.0	13.4	14.4	16.2	18.2	19.8
50th percentile	27.8	27.9	25.4	22.1	18.4	16.0	14.9	15.9	17.9	20.8	23.5	25.5
86th percentile	35.2	34.4	31.1	27.2	21.6	18.0	16.7	18.9	22.3	26.8	30.7	33.2
Rainy days	5	4	6	9	14	13	16	16	13	11	8	7
BRISBANE												
14th percentile	27.2	27.0	26.3	24.6	21.4	18.9	18.4	19.7	21.6	23.3	24.9	26.4
50th percentile	29.1	289.0	28.2	26.5	23.5	21.2	20.6	21.8	23.8	25.7	27.4	28.8
86th percentile	31.1	30.9	30.0	28.4	25.6	23.3	22.7	23.8	26.1	28.2	29.7	31.0
Rainy days	14	14	15	11	10	8	7	7	7	10	10	11
SYDNEY												
14th percentile	22.3	22.4	21.9	19.5	17.2	15.1	14.5	15.0	16.3	17.8	19.6	21.2
50th percentile	26.1	26.1	25.2	22.8	29.9	17.4	16.8	18.0	20.2	22.2	23.9	25.6
86th percentile	29.7	29.5	28.4	26.1	22.5	19.8	19.3	21.1	24.3	26.9	28.6	30.3
Rainy days	11	11	12	11	11	11	9	10	10	11	11	11
MELBOURNE												
14th percentile	19.7	20.0	18.5	16.1	13.5	11.7	11.1	12.1	13.3	14.9	16.7	18.4
50th percentile	25.8	25.6	23.8	20.2	16.6	13.9	13.3	14.8	17.1	19.5	21.8	24.1
86th percentile	33.9	33.3	30.3	25.3	19.9	16.3	15.6	17.7	21.2	25.1	28.3	31.5
Rainy days	8	7	9	12	14	14	15	16	15	14	12	11

Worksheet 4: Box-plots

This page uses temperature data for Perth and Darwin to show the process of drawing box-plots.

January average maximum temperatures

Date	Raw data Perth	Raw data Darwin	Ordered data Perth		Ordered data Darwin	
1	25.7	29.1	22.9	⟷	24.1	Minimums
2	28.9	27.2	23.4		27.2	
3	27.5	24.1	24.0		27.3	
4	24.7	28.4	24.1		28.1	
5	27.6	30.7	24.2		28.4	
6	32.4	32.6	24.4		29.1	
7	24.2	32.2	24.4		29.1	
8	23.4	32.4	24.4	– – –	29.1	Lower quartiles
9	24.0	32.4	24.5		30.1	(24.45 and 29.6)
10	24.5	31.9	24.7		30.7	
11	28.7	29.1	25.7		30.8	
12	38.0	32.0	26.1		31.4	
13	24.4	30.8	27.0		31.5	
14	24.1	32.5	27.5		31.5	
15	27.0	33.8	27.6		31.8	
16	32.9	32.6	28.7	⟷	31.9	Medians
17	35.1	32.6	28.9		32.0	
18	35.1	32.8	29.4		32.1	
19	39.1	32.1	29.7		32.1	
20	41.0	31.5	30.4		32.2	
21	26.1	33.5	30.5		32.4	
22	29.4	31.4	32.4		32.4	
23	29.7	28.1	32.9	– – –	32.5	Upper quartiles
24	33.3	27.3	33.3		32.6	(33.1 and 32.55)
25	24.4	29.1	35.1		32.6	
26	22.9	30.1	35.1		32.6	
27	24.4	31.8	35.9		32.8	
28	30.4	33.4	38.0		33.4	
29	35.9	32.1	38.0		33.5	
30	38.0	31.5	39.1		33.5	
31	30.5	33.5	41.0	⟷	33.8	Maximums

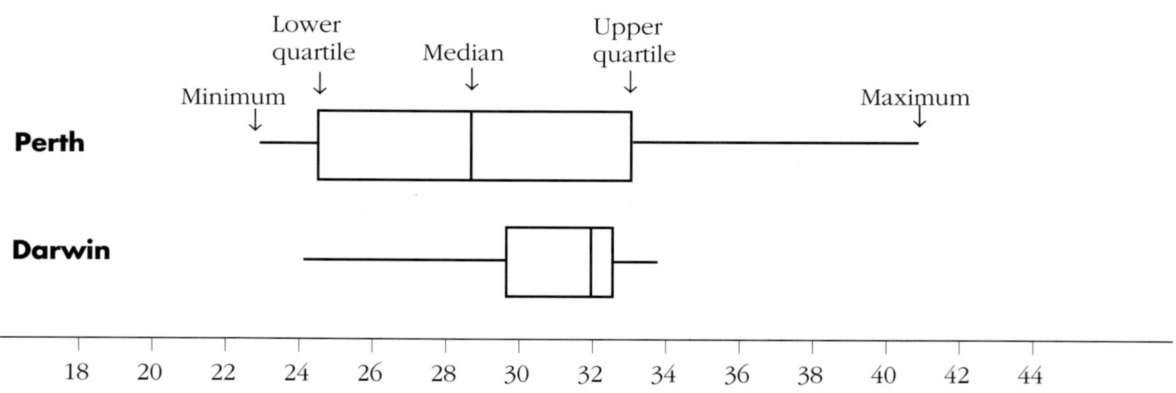

Worksheet 5

Weather patterns in 1990

1 For each month in 1990:

a What were the highest and lowest temperatures in the month?

b What were the mean maximum and minimum daily temperatures?

c Draw either a five-column histogram or a box-plot of the daily maximum temperature.

d On how many days did it rain?

e What was the median amount of daily rainfall?

(Note that, because of the very skewed distribution of rainfall, the Bureau of Meteorology prefers to use the median as the best summary measure of rainfall amounts.)

f Draw either a five-column histogram or a box-plot of the amount of rain each day.

h What was the total rainfall for that month?

2 Describe how the rainfall and temperature changed over the months of 1990 in your chosen city.

3 Describe how the 1990 weather for each month compared with the 'normal' weather for that month. Specifically:

a Compare the 1990 averages to the long-term averages for each month (or for each day if you use the almanac), using mean for temperature and median for rainfall.

b Use the long-term monthly averages and extremes to estimate the percentiles in which the 1990 maximum and minimum temperatures and rainfall for that month occurred.

Cemetery maths

Years 5 to 9

There is something very real about getting data directly from tombstones, finding birth and death dates and how long people lived. The data is no more fascinating than the discoveries that arise about how things have changed over generations. The tool for making these discoveries is statistics, preferably through using a database.

Features

Teaching and learning

- Involves group data collection from a cemetery, either by excursion or research after school hours
- Relates to important social issues, such as community health changes over time, including causes of deaths and patterns of age at death
- Extensive use of a database program to explore the implications of data, including use of graphs
- Relates to life insurance
- Links to history and geography

Major mathematical content

- Using histograms (or stem-and-leaf or box-plots) to compare two sets of data.
- Interpretation of graphs, and searching for relationships
- Interpretation of line graphs of mortality rates
- Using life tables to calculate probabilities including interpolation
- Use of the computer programs: database and spreadsheet

Mathematical content is further explored in the *Content epilogue*.

Equipment and preparation

- It is essential to visit a cemetery or use a listing of data from one (Worksheets 1 and 2)
- A computer database program is desirable to enable analysis of data to move swiftly and easily
- The data on Worksheets 1 and 2 is also on the software disk as a text file

Expected timing

- About 2 weeks

A teacher's story

I got this idea from a social studies book, and from the MCTP activity *Mortality quiz*. It seemed a great way to involve students in real accessible data, where the investigations led to conclusions with social implications and some sound mathematics.

Our local cemetery is near the school, and quite old (in Australian terms). The churchyard contains the graves of a number of local early settlers, including First Fleet settlers. I wanted to do a data-handling unit with a Year 8 class, something different from the 'normal' school surveys. The cemetery proved to be a wonderful alternative resource.

Before taking the class to collect the data, I visited the cemetery and mentally partitioned it into areas containing roughly equal numbers of graves. I divided the class into two groups and arranged for a colleague to accompany me. I decided to collect five pieces of information: Christian name, surname, date of birth, date of death and age at death. I also prepared some 'gravestones' (on grey card) for the students to use to collect the information.

I suggested the idea of exploring data from the local cemetery to students and they were very receptive. I found that I had to use the suggestions written here only as backups for a few groups lacking in ideas.

I negotiated permission for my students to visit the cemetery; I have heard that some cemetery trusts do not wish to have students roaming around but will provide information about the people buried there. Check first; do not leave it to the children.

The students entered their data in the computer database, and we used it to investigate some questions of interest. At this point some fascinating issues were raised. Parents were also intrigued and there was much discussion at home about what the mathematics were beginning to show.

Once we had analysed the local data I introduced the *life tables*, produced by the Australian Bureau of Statistics, which provide a national summary of how long males and females live and hence the death rates for each age. These raised many interesting social issues, particularly the dramatic rise in the death rate of males in their early 20s.

The tables are available as a computer program, so that we were able to draw line graphs, histograms, etc. to compare national data with local evidence.

Finally I showed them how to use the tables to calculate the probability of being alive at any given time in the future, and explained how this helps life insurance companies calculate their premiums.

Comments

This series of lesson was very worthwhile for my heterogeneous Year 8 class. Although a number are not normally keen on maths, they all enjoyed the activity and learned a lot from it. As well as data handling skills, it developed a sense of purpose and appreciation of mathematics as a tool.

At one school, three teachers (maths, social science and English) made the excursion cross-disciplinary. Each teacher produced worksheets and took one third of the group. After an hour the children swapped groups. Having a parent or two would have helped even more.

Summary of the stages in this investigation

1 Posing the issues

2 Collecting the data

3 Entering the data into the database

4 Exploring the data to find answers to questions

5 The life tables: what they tell us

6 Comparing national and local data

7 Using life tables to estimate chances of being alive

1 Posing the issues

I found a great deal of interest generated merely by suggesting the collection of data from the cemetery. Some students told us about their relatives who were buried there, others about what they had found there. There is a fascination with cemeteries that makes this a great experience for students.

If your local cemetery is not so old, consider getting secondhand data from a nearby older one, and taking students on an excursion there. They could roam around finding answers to questions like:
- Who is the oldest (or youngest) person here?
- Which is the oldest burial?
- How many people are buried here?
- Which tombstone has the most (or least) words?

I try to let students come up with the ideas. In case they cannot think of anything to investigate, some suggestions are listed.

Try finding answers to these questions:

1. **Do women or men live longer? (Was the oldest person to die male or female?)**
2. **How has the age at death changed over the years?**
3. **In which age group do more people die? (How many people lived to over 80?)**
4. **In which month do most/least people die? (Do more people die in winter than summer?)**

Answer any other questions of interest to you. For example, what can you find out about causes of death?

2 Collecting the data

Again I tried to get students to decide what information they needed to collect. I stressed that they could not easily go back to get more information later. Here are the suggestions I had in reserve.

I had prepared a number of grey cards, shaped like 'tombstones', for students to use. I initially gave out two for them to use. They had to return these two cards completed and were then given several more.

Students were asked to bring only a pen or pencil. Because the local cemetery is near the school I took the class for a walk. They were organised into their groups reporting either to me or the other adult. At the churchyard, the groups split up, children working in pairs to cover a small section of the churchyard.

Working in this way, with the short walk to the churchyard and back, a 50 minute lesson allowed us to collect enough data to be useful, although we didn't collect all the information available. We all tried to fill in as much as possible.

Collect at least the following information about each person buried or cremated in this cemetery:

first name, surname, sex, birth date, death date, age at death.

Giving the students only two cards initially allowed me to check that they were filling them in correctly.

It takes quite a while for a class of students to gather all the information they need from hundreds of tombstones.

Some problems encountered were:
- difficulty with reading the carved gravestones because of weathering and the old style script;
- difficulty identifying Christian and surnames, especially on multiple graves where family name was often separate from Christian name;
- students straying out of their given area and copying the same information as another pair.

These were not major problems and with more adult help would probably be solved. They did allow us to discuss later the necessity for being organised when data was being collected, the difficulty of collecting data and the need to copy correctly.

3 Entering the data into the database

Our collection of cards was already a 'database' but we agreed it would be easier to analyse if the data were tabulated. It took a number of periods for students to enter all the information into the computer database since each student added only a few records per session.

We created the following fields:
- first name
- surname
- sex
- birth date
- death date
- age at death

They entered each record with M or F, as appropriate, and the date.

You may need to check on the way the program requires the date to be entered, for example, in the form 12jan1899.

Some tombstones had missing data, so what could not be inferred was entered as Ø.

4 Exploring the data to find answers to questions

As much as possible, I arranged to let small groups of students have access to the computer for about 10 minutes at a time, while others were doing something else. Before going to the computer, they should have drawn up a plan of action to use to answer their own questions or the ones given in stage 1.

The following refers to the questions from stage 1.

Questions 1 and 3

Compare the histograms (or box-plots, or stem-and-leaf plots) and the statistics for the 'age at death' of two groups: males and females. This will show the shape of the distribution and you can see which age group is least and which is most commonly represented. Are there any surprises? Can you explain them?

The statistics will also give the average (mean) age of death for men and women over the period that the cemetery has been in use. Are there many young children in the sample? If so, have they affected the average? Is the median a better indicator of average?

Question 2

Find the median of the 'dates of birth'. Use this to split the sample into two parts: those born before it, and those born after it. Compare male with female 'age at death' for both the 'before' and 'after' sets of people. Describe any differences. Try to explain them.

If you have a large enough sample, make ten-year divisions according to 'date of birth'. In which decade were there most deaths? For each decade compare the number of deaths and the average age at death of males and females. Is there a trend?

This was our first go at using a computer database in a maths class. We shared our successes and the occasional frustration.

It was great how all the statistics we had learned were found to be useful in analysing the data.

I found it more flexible to provide hard copy of the data as well as having them in the computer database.

Extension

Worksheets 1 and 2 give data from a cemetery in Victoria and one in Tasmania. In addition there are a number of commercial databases (for example *First fleet*) which provide data on ages at deaths of populations. These could be explored for comparisons with the local data collected.

5 The life tables: what they tell us

Using official data from the Australian Bureau of Statistics (ABS), the insurance industry publishes life tables. These take an imaginary 100 000 people, and give the number that can be expected to be alive at various ages. These numbers are different for males and females. They are simplified on page 396.

I copied the table on page 396 for the students. They drew graphs like the computer-drawn graph (to age 50 only) shown here. Using the table and appropriate graphs it is possible to work out many calculations.

The full version of life tables is found in *Deaths, Australia* (latest year) from the Australian Bureau of Statistics.

Until what age will half the males still be alive?

Until what age will half the females still be alive?

What is the safest age to be? At what age do you have the smallest chance of dying in the next year?

At what age do any peaks occur? Why?

At what ages are the low points on the graphs? Why?

At what age are one tenth of the people expected to die in the next year?

ABS life data 1990

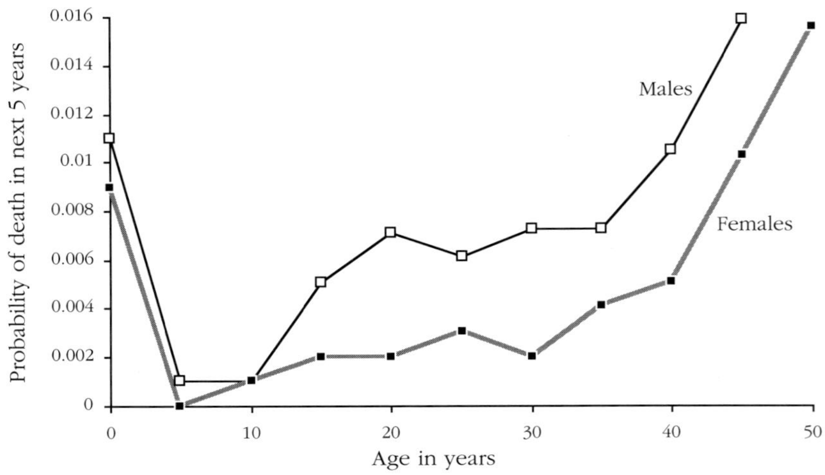

6 Comparing national and local data

The software disk may include a file called *Life tables*, which is data for 1990. Students can use it to draw:

- a line graph of the number of people alive at any age, for males and females;
- a histogram of the age at death of the general population of males and females;
- a line graph showing the death rate at different ages.

The information in *life tables* can be compared with the local data; it should be remembered that the local data relate to the total period over which the cemetery has been used, while life tables refer to 1990.

To get the death rate you simply divide the number who are expected to die at any age by the number alive at the start of that year. This can be used to check the answers to the questions above.

Number of original 1000 people alive at different ages

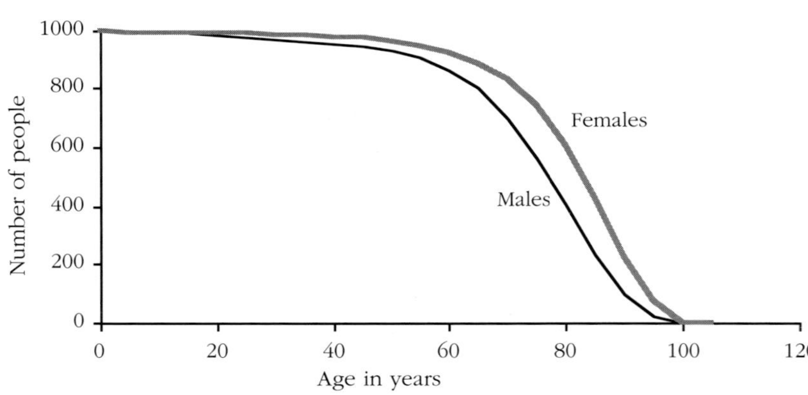

7 Using life tables to estimate chances of being alive

The life tables on page 396 give the number of people alive every 5 years. So we can use them to compare the numbers still alive at any age in the form of fractions or percentages. This fraction (or percentage) can be interpreted as the probability that any particular person will still be alive at that age.

The computer program *Cemetery maths,* if available, uses the same data to compute these fractions.

I raised the question of how a life insurance company would be able to use these data to decide how much it would have to pay out each year, and hence what annual premiums to charge.

Example

Number of males born:

A = 100 000

Number of males alive at age 20:

B = 98 210

Fraction of males alive at age 20:

$A \div B = 0.982$

This means that the probability that a male baby will still be alive at age 20 is 98%.

The same method can be used to find the probability that a person of any age A will still be alive at age B.

Content epilogue: background information for teachers

Comparing two sets of data

Many of the questions that students will wish to answer will require comparing two sets of data on the same item of interest (for example, males and females, or adults and children, or before 1950 with after 1950).

Once the data has been obtained, the best way to make a comparison is to use a graph that shows both sets of data at the same time, but distinguishes one from the other. Two useful methods are illustrated here, namely side-by-side histograms and back-to-back stem-and-leaf plots, showing age at death.

Side-by-side histograms

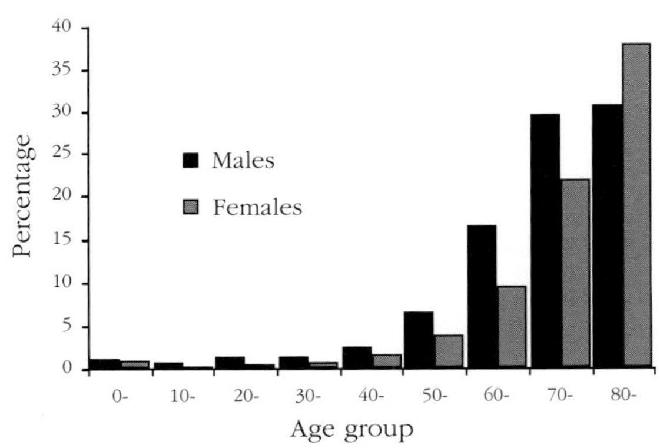

Back-to-back stem-and-leaf plots

Males		Females
1 4 0	8	3 4 7 8 2
6 3 2	7	2 5 8 3 1 0
9 7 5 2	6	0 0 1 1 2 4 7
9 5 3 2 0	5	3 5 0
9 8 4 4 7	4	3 3 0
9 9 8 3 1	3	5
9 4 5 7 1	2	0
9 9 8 7	1	

The *stem* is the tens digit and the *leaves* are the units. The top line on the males side shows three ages (81, 84 and 80); whereas there are five entries on the top line on the female side (83, 84, 87, 88 and 82).

Relationships and scatter-plots

Do 'ages at death' increase as time progresses? You could seek to find out if we are living longer by comparing ages at death with the year in this century. This method can be used to see if any two sets of values could be related in some way. Again, the best thing to do is to organise the data well and draw a suitable graph.

One way to display this data is to use coordinates; it is called a scatter-plot. An example is shown.

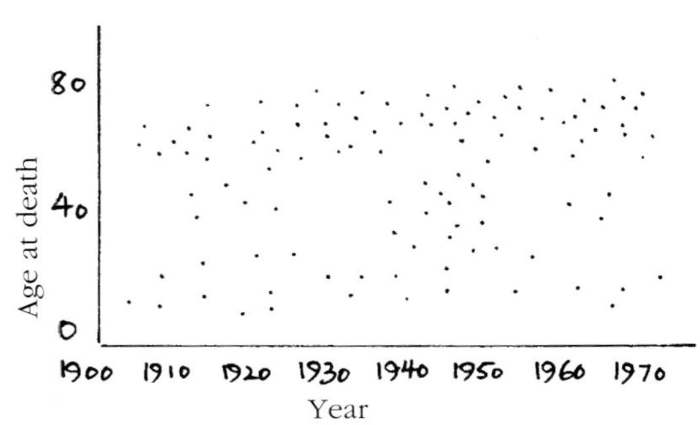

If there is a relationship that shows we are living longer, then the points should cluster around a line that slopes upwards. The relationship may not be very exact; if it is not clear that the points cluster at all, then it is not sensible to infer any relationship. (A mathematical measure of the 'amount of clustering' is called 'correlation'. There are mathematical techniques to find the 'best fit' of a straight line to the cluster of points; they are studied in statistics at senior secondary levels.)

Life tables

The tables as published by ABS give the results of many calculations. However, the ones of greatest interest are illustrated by means of the 5-year age groups below. It is suggested that you use a calculator to check that you understand how the numbers are obtained.

- The number dying each 5 years is just the result of a subtraction.
- The death rate, or probability of death, is the number dying in the 5-year period divided by the number alive at the start of the period.
- The probability of living from age C to age D is just the fraction of those alive at age C who are still alive at age D (or 1 – the probability of dying).

ABS life tables 1990 (simplified)

Age	0	5	10	15	20	25	30	35	40	45	50	55	60	65	70	75	80	85	90
Number still alive at each age																			
M	1000	989	988	987	982	975	969	962	955	945	930	906	864	799	699	566	403	235	97
F	1000	991	991	990	988	986	983	981	977	972	962	947	924	887	829	740	609	430	229
Deaths																			
M	0	11	12	13	18	25	31	38	45	55	70	94	136	201	301	434	597	765	903
F	0	9	9	10	12	14	17	19	23	28	38	53	76	113	171	260	391	570	771
Number of deaths in next 5 years																			
M	11	1	1	5	7	6	7	7	10	15	24	42	65	100	133	163	168	138	73
F	9	0	1	2	2	3	2	4	5	10	15	23	37	58	89	131	179	201	154
Probability of death in next 5 years																			
M	0.011	0.001	0.001	0.005	0.007	0.006	0.007	0.007	0.010	0.016	0.026	0.046	0.075	0.125	0.190	0.288	0.417	0.587	0.753
F	0.009	0.000	0.001	0.002	0.002	0.003	0.002	0.004	0.005	0.010	0.016	0.024	0.040	0.065	0.107	0.177	0.294	0.467	0.672

The top table of these four could be copied for students to use at stage 5. The data for any age can be interpreted by *interpolation*; that is, working out what happens within each group of 5 years.

Data from cemeteries at Epping and Rokeby

The data from two cemeteries are included in Worksheets 1 and 2. They are also on the software disk so that they may be entered into a database and compared with your own data. The date of death is given as a month and year. Epping is north-east of Melbourne.

The township of Rokeby lies on the eastern shore of the Derwent River, near Hobart. Once called Clarence Plains, it was settled as a farming area. In 1808 a number of men from Norfolk Island were given grants of land in the area. The gravestones of some of the people can be found in the churchyard. Two of these men, Edward Kimberley and James Morrisby, are included on the database.

Many early pioneer families are buried in the churchyard. Probably the largest group in the area was the Morrisby family. They are descendants of James Morrisby who arrived on the First Fleet and is buried in the churchyard. The church is still in use, as is the graveyard, as indicated by the burials up to 1992.

Worksheet 1: Data from Epping cemetery, Victoria

First name	Surname	Sex	Born	Age at death	Date of death		Place
Thomas	Ahern	M	1819	46	3	1865	Epping
Bridget	Ahern	F	1829	79	7	1908	Epping
Bridget	Ahern	F	1906	6 days	7	1906	Epping
Stephen	Brazil	M	1854	83	10	1937	Epping
James	Brazil	M	1881	67		1948	Epping
Agnes	Brazil	F	1889	102		1991	Epping
Daniel	Breen	M	1837	35		1872	Epping
Patrick	Breen	M	1843	66		1909	Epping
Obias	Butler	M	1828	43	9	1871	Epping
George	Byrne	M	1853	82	10	1935	Epping
Patrick	Clarkes	M	1873	60	11	1933	Epping
Joseph	Cleeland	M	1830	88	9	1918	Epping
Mary	Cleeland	F	1842	75	4	1917	Epping
Patrick	Clune	M	1869	40	3	1909	Epping
Bernard	Corrigan	M	1871	50	3	1921	Epping
Elizabeth	Corrigan	F	1877	91	5	1968	Epping
George	Davis	M	1878	64	6	1942	Epping
Hugh	Dunganmo	M	1882	31	12	1913	Epping
Agnes	Flemins	F	1860	72	5	1932	Epping
John	Flemins	M	1862	83	6	1945	Epping
Harry	Goodmen	M	1869	85	1	1954	Epping
Mary	Goodmen	F	1870	81	1	1951	Epping
Richard	Goss	M	1860	84	10	1944	Epping
Ellen	Goss	F	1866	68	2	1934	Epping
Bernard	Gross	M	1834	74	5	1908	Epping
Margaret	Hanuske	F	1887	68	2	1955	Epping
Fred	Henry	M	1808	58	8	1866	Epping
Leslie	James	M	1912	22	12	1934	Epping
Eleanor	Jane	F	1863	82	1	1945	Epping
Billy	Jolly	M	1941	3yrs 11 mths	2	1945	Epping
Anne-Marie	Kerran	F	1881	80	3	1961	Epping
James	Long	M	1876	68		1944	Epping
Margaret	Long	F	1882	65		1947	Epping
John	Long	M	1920	36		1956	Epping
William	Louden	M	1863	2yrs 8 mths	6	1866	Epping
Alice	Louden	F	1865	11 mths	5	1866	Epping
Liz	Macey	F	1873	81	2	1954	Epping
Elizabeth	Macey	F	1873	81	2	1954	Epping
Ryan	Macey	M	1910	64	8	1974	Epping
Hans	Maurer	M	1856	86	4	1942	Epping
Eliza	Maurer	F	1862	76	12	1938	Epping
Patrick	McGrath	M	1878	72	6	1950	Epping
Eliza	McGrath	F	1883	80	10	1963	Epping
Timothy	Murphy	M	1887	81	7	1968	Epping
Terrence	O'Hern	M	1814	90	6	1904	Epping
Margaret	O'Hern	F	1818	86	7	1904	Epping
Johanna	O'Sullivan	F	1784	99	9	1883	Epping
Daniel	O'Sullivan	M	1820	80	7	1900	Epping
Margaret	O'Sullivan	F	1840	24	7	1864	Epping
Eugene	O'Sullivan	M	1863	7 mths	1	1864	Epping
Gertrude	Olsen	F	1887	5.5	3	1892	Epping
Elizabeth	Pratt	F	1889	64	2	1953	Epping
Reginald	Sims	M	1890	28	8	1918	Epping
Maria	Timm	F	1807	83	12	1890	Epping
Jane	Tulloch	F	1832	32	10	1864	Epping
Alexander	Tulloch	M	1858	27	1	1885	Epping
John	Tulloch	M	1863	34	1	1897	Epping
Denis	Tuttle	M	1874	78	5	1952	Epping
Mary	Tuttle	F	1905	52	12	1957	Epping
Margaret	Tuttle	F	1909	11 mths	2	1910	Epping
John Henry	Withell	M	1885	45	5	1930	Epping
Prudence	Withell	F	1888	43	6	1931	Epping
John Thomas	Withell	M	1911	18	10	1929	Epping

Worksheet 2: Data from Rokeby cemetery, Tasmania

First name	Surname	Sex	Born	Age at death	Date of death		Place
Fred	Abbott	M	1862	82	10	1944	Rokeby
Lillian	Abbott	F	1872	86	10	1958	Rokeby
Arthur	Atkinson	M	1906	73		1979	Rokeby
Grace	Badman	F	1892	26	5	1918	Rokeby
Emma	Burton	F	1862	28	12	1890	Rokeby
Mary	Calvert	F	1891	84	4	1975	Rokeby
Margaret	Chipman	F	1869	54	11	1923	Rokeby
John	Clement	M	1899	38	4	1937	Rokeby
William	Dart	M	1822	58	11	1880	Rokeby
Mary	Easy	F	1851	65	12	1916	Rokeby
Margaret	Elizabeth	F	1851	85	10	1936	Rokeby
Frederick	Fazackerley	M	1860	54		1920	Rokeby
Alice	Garlick	F	1912	67		1979	Rokeby
Sylvia	Gorringe	F	1914	62		1976	Rokeby
Laura	Groves	F	1910	77	8	1987	Rokeby
Violet	Inslow	F	1907	60	7	1967	Rokeby
Laurence	Irving	M	1890	60		1950	Rokeby
Rupert	Isles	M	1916	41	4	1957	Rokeby
Elizabeth	Johnston	F	1892	86	6	1978	Rokeby
George	Johnston	M	1916	70	8	1986	Rokeby
Henry	Jollife	M	1850	65	7	1915	Rokeby
James	Joseph	M	1794	36	5	1830	Rokeby
Benjamin	Joseph	M	1817	95	7	1912	Rokeby
Margaret	Joseph	F	1819	74	5	1893	Rokeby
John	Joseph	M	1856	81	9	1937	Rokeby
Edward	Kimberley	M	1743	67	11	1829	Rokeby
Jane	King	F	1829	42	3	1871	Rokeby
Robert	Knopwood	M	1760	78	9	1838	Rokeby
William	Lamont	M	1893	46	5	1939	Rokeby
Richard	Larson	M	1792	57	7	1849	Rokeby
Dahlia	Lazenby	F	1889	20	11	1909	Rokeby
Willis	Lazenby	M	1905	50	8	1955	Rokeby
George	Massie	M	1901	71	6	1972	Rokeby
Charlotte	Massie	F	1901	84		1985	Rokeby
Peter	Matthews	M	1884	21		1909	Rokeby
Mark	McGrath	M	1972	11	1	1983	Rokeby
Linda	Millhouse	F	1903	84	9	1987	Rokeby
Janice	Millhouse	F	1945	43	9	1988	Rokeby
Herbert	Moorhouse	M	1899	91	7	1990	Rokeby
Lissy	Moorhouse	F	1902	75	10	1977	Rokeby
John	Morrisby	M	1831	92	9	1923	Rokeby
Raymond	Morrisby	F	1878	56		1934	Rokeby
Keith	Morrisby	M	1893	66	12	1959	Rokeby
Rosina	Morrisby	F	1894	25	1	1919	Rokeby
Helen	Morrisby	F	1924	5	4	1929	Rokeby
Yvonne	Morrisby	F	1954	33		1987	Rokeby
Albert	Morrisby	M	1965	21		1987	Rokeby
James	O'May	M	1845	89	11	1934	Rokeby
Mary	O'May	F	1851	85	10	1936	Rokeby
Alice	O'May	F	1875	70	5	1936	Rokeby
Charles	Owen	M	1868	87	5	1955	Rokeby
Anne	Pedder	F	1820	75	7	1895	Rokeby
Stuart	Pizer	M	1946	40	12	1986	Rokeby
Sybil	Rothlisberger	F	1942	50	1	1992	Rokeby
Rosina	Shuttle	F	1809	3	12	1812	Rokeby
Terence	Since	M	1879	73		1952	Rokeby
Emily	Spratt	F	1863	101		1964	Rokeby
Alan	Steel	M	1948	40	2	1988	Rokeby
Geoffry	Tollard	M	1934	50		1984	Rokeby
James	Walter	M	1855	80	7	1935	Rokeby
Anne	Wilcock	F	1790	64		1854	Rokeby
William	Winspear	M	1852	53	6	1905	Rokeby
Archibald	Wize	M	1905	76		1981	Rokeby
Margaret	Young	F	1863	64	5	1927	Rokeby
Edwin	Young	M	1878	64	4	1942	Rokeby

8 Sampling and inference

The four investigations in this chapter focus on the use of samples to provide information about a larger 'population' from which the sample comes. This is a very important aspect of statistics in the real world, and is the mathematics behind opinion polls, market research, etc. It is in this topic that statistics and probability (chance and data) come together most firmly. In fact many of the other investigations in this book also involve sampling as part of that process.

The four investigations involve making informed guesses about a larger population based on the properties of a sample. For example, *What's in the bag?* uses a guessing game in which the mystery contents of a bag are to be guessed from the clues given — five samples drawn from the bag and replaced. In *Duelling dice* students explore, through a game, the strange relationship between four special dice. This involves sampling in that the rolls are random. Later a theoretical analysis provides information about the 'population'.

How long is a rod? and *Colas* explore quite deep and significant ideas. *How long is a rod?* shows that averaging a large sample will give a better estimate of a population mean than averaging a small sample: it leads to some understanding of the Central Limit Theorem. *Colas* develops into an exploration of the binomial probability involved in random guessing, and asks whether a 'real' percentage of correct answers is likely or not to represent guessing. This is the start of significance testing, hypothesis testing and confidence intervals.

From *A National Statement on Mathematics for Australian Schools*

(Overview) The dual notions of sampling and of making inferences about populations, based on samples, are fundamental to prediction and decision making in many aspects of life. Students will need a great many experiences to enable them to understand principles underlying sampling and statistical inference and the important distinctions between a population and a sample, a parameter and an estimate... the groundwork will be laid in the early years of schooling in the context of data handling and chance activities.

(Band B) During these years, children will begin to make informal inferences on the basis of samples and discuss whether it is possible to generalise... They should be encouraged to question the appropriateness of generalisations beyond the bounds of the data collected and begin to consider the circumstances under which generalisations may reasonably be made.

(Band D) All students should develop at least an informal understanding of confidence intervals, even if a more technical treatment of this is inappropriate. Confidence intervals can also be used as an informal approach to hypothesis testing. (pp. 164, 168, 181)

What's in the bag?

Years 2 to 8

Out of an apparently simple game an extraordinary number of mathematical features can develop. Students take a sample from a bag of coloured pegs. On the basis of their sample, they guess what's in the bag. By taking extra samples they change their prediction and increase their confidence. Finally the contents of the bag are disclosed.

Features

Teaching and learning

- Generating firsthand data for analysis
- Problem solving
- A guided investigation
- Concrete materials
- Group work
- Estimation and prediction
- Language
- Computer simulation

Major mathematical content

- Finding and recording outcomes
- Representing data
- Using the language of chance
- The notion of sampling to predict a population
- Using secondhand data
- Combines probability and data handling

Mathematical content is further discussed in the *Content epilogue.*

Equipment and preparation

- Coloured pegs or cubes (e.g. Unifix) in sufficient colours and number for each group to be able to select a sample of ten pegs
- Empty paper bags to hold the pegs or cubes so they are not visible when samples are withdrawn
- Check the operation of the available computer program

Expected timing

- Generally two sessions, though students may want to continue the game over several weeks

A teacher's story

I have noticed that young students, in particular, and some adults are prone to generalise about a population based on remarkably little evidence. With this in mind, I thought a guessing game, where students keep getting more information in the form of samples enabling them to refine their guesses, would help them take the important statistical step of attaching a confidence level to their estimates.

Children are challenged with predicting the colours of a given number of pegs (or cubes) in a paper bag. A sample of pegs is withdrawn and the children predict the colour of all the pegs, including those still in the container. The sample is then replaced and the colours thoroughly mixed up. Another sample is drawn and the children now revise their predictions. This process is repeated a number of times with the children's confidence in their revised predictions improving each time.

The computer program gave me the chance to extend the activity as far as the children's interests and abilities allowed. In addition, the game associated with Worksheet 2 allowed the children to work with and analyse secondhand data.

Comments

The investigation covers the concepts of sampling and elementary probability. The intuitive analysis and prediction is a valuable precursor to the more formal approaches to probability.

Feedback is available as a reinforcement for predictions. The concepts of probability are steadily constructed by students in a non-threatening way.

Children loved the guessing aspect, and the anticipation just before the contents were revealed was great.

They wanted to continue the sampling and guessing until they were logically certain they were right. They wanted to turn it into a 'Mastermind' type game.

Summary of the stages in this investigation

1 Posing the problem

2 Predicting what's in the bag

3 Assessing the reliability of predictions

4 Making and playing their own games

5 Computer simulation

6 Consolidation

1 Posing the problem

I wanted to illustrate the difference between guessing, knowing and inferring (guessing based on evidence), so I put six pegs (using four different colours) into a paper bag. I didn't tell the children anything except that there were pegs of different colours in the bag.

Guess how many of each colour in the bag.

Clearly their guesses were completly 'wild', so I said 'Would you like a clue?'.

They said 'Yes', so I tipped out *all the pegs* for them to count.

We agreed this was a rather 'big' clue.

I then made up another bag also with six pegs.

I will give you a clue and show you *five* of them. What colour do you think is left?

They could readily see they were simply guessing about the one peg they had not seen. I mixed up the pegs and took out another five; they looked at these very closely to see if there was an extra clue.

What about another lot of five? Are there any more clues?

The purpose of the game we are about to play is to make good guesses based on the clues you have.

2 Predicting what's in the bag

To introduce the structure of the game, I created a mixed set of eight pegs in four colours in the bag. I told them that much.

I will let you look at a sample of four pegs at a time. This is one clue. You can have four more clues.

I handed out Worksheet 1. After each sample of four pegs was shown, students wrote down what they had seen.

After each sample I put the pegs back and shook up the bag. I repeated sampling four times and then asked for their predictions.

	Sample	Prediction
First	R R B B	R R B B B Y G G (or 2R, 3B, 1Y, 2G)

I learned the value of silence. Once we started sampling, I deliberately did not say anything. It led to the children being very quiet too, and I could see them all thinking hard.

3 Assessing the reliability of predictions

I now asked how certain the students were of the accuracy of their prediction.

What did your neighbour guess? Have a short discussion, change your guess if you would like to.

> They were peering at their neighbours' guesses anyway, so it was a good idea to formalise this sharing.

Most were pretty sure by now and wanted to know the 'right answer'. So with a bit of drama, I let eight students come out one by one and take out a peg. There was a fair bit of anticipation for the last peg.

We discussed whether the drawing of more samples or larger samples would improve their confidence in their answers.

> They were all very close with many exactly correct. One student was a bit upset when he noticed his very first guess (without clues) was exact and then he changed his mind.

4 Making and playing their own games

With the children organised into groups I invited one group member to put any ten pegs into a bag.

To play the game, another group member takes a sample of four pegs and shows them to the rest of the group. They all record the colours of the sample and predict the colours of the ten pegs on the basis of the sample.

The sample is returned to the bag and, after shaking, a second person takes a sample of four. All group members again record the colours and revise their predictions of the total set of ten pegs. This is repeated two or three times. Finally the ten pegs are shown and everyone's final predictions of the colours are checked.

For a new game a different person selects the ten pegs.

> I made the activity into a competition with the child who predicts the greatest number of correct pegs (or predicts them first) getting to choose the ten peg colours for the next game.
>
> I found the game provided me with a marvellous opportunity to assess their level of understanding of the chance process. Some children devised colour choices that made it difficult for others to predict. For example four red, five blue and one yellow. (They found that the yellow rarely featured in any of the drawn samples.)

5 Computer simulation

There are three distinct parts to the program (if available to support the lesson).

Demonstrate sampling

- The computer creates and displays the contents of a bag, up to 20 pegs in four colours.
- The user chooses a sample size, say five.
- Then, *knowing the contents* of the bag, the students guess what will come out.
- The computer reveals the sample.

The major purpose of this initial demonstration is to build a trust that the simulation does exactly what the students do with the real objects.

What's in the bag?

This time the contents of the bag are hidden. The program provides samples of a chosen size. This continues until the user requests the correct answer.

There are two options:

1 Students enter their guesses, and the computer tells them when they are correct.

2 Students receive a print-out of the samples. They then return to their desks to analyse the data before returning to the computer with a prediction.

The computer 'guesses' the contents of a bag

There are two ways to use this alternative program:

1 Students physically construct a bag and then provide samples of information for the computer which updates its prediction. One use of this program is to compare the accuracy of the computer to the accuracy of students.

2 The computer makes up and displays the contents of a bag. But then it takes samples from this bag as if it does not know the contents and displays its prediction. Students can watch and record how many guesses it needs to find the correct combination. Records are kept which display the average number of guesses needed.

6 Consolidation

The following class games provide opportunities to practise the skill of estimating based on clues offered through samples.

Daily 'quick predict'

I made up a large container of 100 pegs, with lots of colours. Twice a day (at the start and end) we drew out a sample of 10, and recorded it on a chart next to the container. On Friday, everyone made their final guess and we tipped out the container — the closest few were rewarded with a little prize.

'Quick predict' using the computer or Worksheet 2

Using Worksheet 2, I cut a strip of five samples for each group, and ran the activity in the same way as the following suggestions for the computer.

Each group in the class came out to the computer and received a print-out of five samples (I chose 20 pegs, four colours and a sample size of five.) Then each group had a few minutes to predict the contents. (There was much eager discussion in this time.) Each group's prediction was written on the board. We noted variations in the predictions and then asked the computer for the correct answer.

A points scoring system helped create a friendly sense of competition. We had five groups, so when the answer was written on the board, the group who were closest received 6 points, the next got 4, then 3, then 2, then 1. We played five games, each game taking about 10 minutes to distribute the samples, have the discussion and prediction, get the answer and allocate the points.

There was some fascinating mathematics involved in deciding who was closest and in allocating points where there was a tie.

'Quick predict', against the computer

We had a class competition against the computer. Using the choice of 20 pegs, four colours and samples of five, we played five games with the students predicting, and found how many samples we needed to get the right combination. Then we used the program to set the computer the same challenge. We ran it slowly (sample by sample) to compare progress. The students won four of the five games and felt very good about themselves.

The computer program can be used to explore how the ability to correctly predict the contents of the bag varies with the size of each sample. With this option the average number of samples that need to be drawn to correctly identify the contents are recorded for samples of varying sizes.

Answers to problems on Worksheet 2:

1 12R, 3B, 3G, 2Y

2 3R, 7B, 2G, 8Y

3 7R, 5B, 3G, 5Y

Through this activity children gained an understanding of the value of a computer simulation where parameters can be varied and the consequences determined easily and quickly.

Content epilogue: background information for teachers

Samples as predictors of a population

There are many notions of *sampling and probability* embedded in this investigation. Perhaps these are best explained by describing the routine the computer uses to predict the contents of a bag.

Say the bag contains 20 pegs of four colours and we take samples of five.

Let us suppose the bag contains *three* red, *seven* blue, *two* yellow and *eight* green.

Sample 1 is R B B Y G

The computer logic: Blue is the most common; there are probably twice as many blue as any other colour, and the others are all equal.

Since two-fifths of the sample is blue, the reasoning goes that two-fifths of the total bag is probably blue. Since two fifths of 20 is eight, the prediction is there are eight blue. And since one fifth of 20 is four, the prediction is four red, eight blue, four yellow, four green.

Sample 2 is G G G B R

The same reasoning as above would produce a prediction of four red, four blue, no yellow and 12 green.

However, combining the two results so far:

Total sample of ten is 2R, 3B, 1Y and 4G

The reasoning now predicts that the 20 pegs are double this: four red, six blue, two yellow and four green.

This prediction is much closer to the actual contents, and our confidence in this guess is much higher than our confidence in the guess after just one sample.

Worksheet 1

Problem 1

	Sample	Prediction
First		
Second		
Third		
Fourth		
Fifth		

Problem 2

	Sample	Prediction
First		
Second		
Third		
Fourth		
Fifth		

Problem 3

	Sample	Prediction
First		
Second		
Third		
Fourth		
Fifth		

Problem 4

	Sample	Prediction
First		
Second		
Third		
Fourth		
Fifth		

Problem 5

	Sample	Prediction
First		
Second		
Third		
Fourth		
Fifth		

Problem 6

	Sample	Prediction
First		
Second		
Third		
Fourth		
Fifth		

Worksheet 2

The following are sets of samples drawn from a bag with 20 pegs and four colours. The colours are red (R), blue (B), green (G) and yellow (Y). Cut up a copy of this sheet so that each group gets one set of samples, and predicts the contents of the bag. A points scoring system is suggested in stage 6. Answers are also included.

Problem 1 Group 1	Problem 1 Group 2	Problem 1 Group 3	Problem 1 Group 4	Problem 1 Group 5	Problem 1 Group 6
RRGBB	RRRBY	RRRGY	RRRRR	RRBBG	RRBGY
RRRRG	RRGGY	RRRBG	RRBGG	RRRRB	RRBYY
RRRRB	RRRRG	RRBGY	RRBBG	RRRGY	RRRGG
RRRRR	RRRYY	RRGGY	RRRRY	RRRBY	RBGGY
RRRGY	RRRBY	RRBGY	RRBGG	RRBBG	RRRBG

Problem 2 Group 1	Problem 2 Group 2	Problem 2 Group 3	Problem 2 Group 4	Problem 2 Group 5	Problem 2 Group 6
BYYYY	RBBBY	RBYYY	RBYYY	RBBYY	BGGYY
GYYYY	BBGYY	BBGYY	BBGYY	RBYYY	RRRYY
BBGYY	BYYYY	BBGYY	BBYYY	RRYYY	BBBBY
RBBGY	BGYYY	RBBYY	RGYYY	RGYYY	BBGYY
BBGYY	BBGYY	RBBBY	BBGYY	RBYYY	RRBGY

Problem 3 Group 1	Problem 3 Group 2	Problem 3 Group 3	Problem 3 Group 4	Problem 3 Group 5	Problem 3 Group 6
RRBGY	RRRGY	RRBGY	RBBBY	RRRYY	RRYYY
RBYYY	RBGGY	RBBGY	RRBGG	RGYYY	RBBYY
RRBYY	RBBGG	RRBGY	RRBGY	RBBGY	RRRYY
RBBYY	RRRRG	RRRRY	RRBBG	RRRBY	RBBGY
RBBYY	RBYYY	RGYYY	RRRBG	RRRRG	BGYYY

Duelling dice

Years 4 to 12

The Red die is better than the Blue die in a game. Blue is better than Black. Black is better than Green. Surely Red is better than Green? But not so with these four special dice. This investigation, featuring game playing, probability analysis and computer simulation, is driven by the counter-intuitive result.

Features

Teaching and learning

- Group game context
- Data for analysis generated firsthand by the class
- Computer simulation
- An open-ended investigation
- Links to a childhood playground game and to sporting contexts

The dice were first written up in *Scientific American*, Dec. 1970, vol. 223, no. 6.

Major mathematical content

- Collecting and interpreting data
- Forming and testing hypotheses
- Calculation of long-run frequencies
- Observing and generalising patterns
- Developing explanations of what happens
- Simulation modelling
- Probability, evidence of natural variability

Mathematical content is further discussed in the *Content epilogue*.

Equipment and preparation

- For each group of three to four students, make one set of four dice (2 cm wooden cubes are good — use sticky circle labels for the numbers)

Numbers to be marked on the sides of the dice:

Red	0, 1, 7, 8, 8, 9	Blue	5, 5, 6, 6, 7, 7
Black	3, 4, 4, 5, 11, 12	Green	1, 2, 3, 9, 10, 11

- Coloured felt pens
- Set up and check the computer program

Expected timing

- Main activity about three periods
- Probably much longer with extensions

A teacher's story

I came across these special dice while reading *Scientific American* years ago. They fascinated me at the time, and I thought they might fascinate my class; how right I was! They also reminded me of the childhood playground game of 'scissors, rocks and paper'. (I was surprised to find this game still survives, and with some recent inclusions!)

I gave the groups a set of the four special dice to use in games against other groups to find the 'best' die. In the games that followed, students put forward and tested many hypotheses, and this all led to the realisation that these were very unusual dice. Challenging them to put the dice into order (an impossible task) got the unusual property out into the open. We then compared the dice with 'scissors, rocks and paper' and found the same property: whichever you choose, there is always one that can beat it. It also seemed to have parallels with the way sporting results sometimes occur.

Although the results we used were empirical (that is generated from the games), their unusual nature created interest in a more theoretical analysis. There was much problem solving as pupils sought to explain why one die was expected to be better than another.

Some students wanted to invent other dice to fit the problem. The computer program became an investigative tool to help them compare two dice quickly to see which would be the likely winner.

A variety of extensions followed. Students were determined to find a winner, so they rolled all four at once, many times! They also had found that, when playing with two dice, one die usually won, but not always. So we launched into an exploration of how the scores were distributed. Lastly, we changed the number of points needed to win the game, to see how that affected the results. A computer simulation took over the hard work.

Comments

The unusual dice kept student interest high. Underline 6 and 9 to make them different.

As groups of students moved around the room challenging others to games, the involvement was extraordinarily high. The 'different' nature of the activity involved all the class, especially those not used to success and enjoyment in a mathematics class. I will certainly use the investigative approach to probability in future.

I had to work hard with my Grade 4 to get them to 'come to grips' with notions of chance and probability; however they enjoyed playing the game!

Summary of the stages in this investigation

1 **Posing the challenge**

2 **Group investigation: collecting data**

3 **Summarising and discussing data**

4 **Scissors, rocks and paper**

5 **Problem solving**

6 **Theoretical analysis**

7 **Computer simulation**

8 **The search for another die**

9 **Discussion and report**

10 **Consolidation**

11 **Extensions**

1 Posing the challenge

I split the class into groups of three and gave them each a set of the four special dice.

I asked them to look at the dice, and write down the numbers on each. I suggested they may like to investigate them.

This was partly to get them familiar with the dice, and partly to make sure they could tell the difference between the 6 and 9.

Which is the best die to use in a competition against the others?

Each group can challenge another to a competition. Each chooses a different colour die. Roll the dice and the highest number scores one point. First to seven wins the game.

As you play more games you will get closer to answering the question: Which die is the best?

2 Group investigation: collecting data

Groups moved around the room challenging other groups to a game.

You've chosen Red, we'll use Green.

7 3 You win!

9 9 It's a tie! Roll again.

That's four points to one so far. First to seven wins the game.

I drew up tables on the board for students to enter the results as they finished each game. To make sure all six possible games were played at least six or seven times, I encouraged some groups to play certain games later in the session.

Clear instructions, possibly summarised on the board, help the children stay on task.

Have you decided which is the best die to use yet?

As the data base began to emerge it was scrutinised by students for clues as to which die was the best. In several times teaching this activity, no-one has ever suspected the 'circular' pattern. Pupils are convinced one die will be 'best'.

The results show surprising variability: sometimes one die will win convincingly, and in the next game it will lose convincingly. Data gathering should continue until the 'answer' is clear.

3 Summarising and discussing data

I added the scores in the tables. Students in groups discussed the results and tried to rank the dice to decide which was the best — trying to make sense of the unusual results.

One student from each group was selected by the group to briefly report their discussion to the whole class.

It was clear that some students had realised that there was something a bit special about these dice. They found it difficult to say what it was, or to explain it.

I asked them to rank the dice like a football ladder. But they could not, since any attempt to do so has the bottom team able to beat the top team!

After some discussion and seeking pupils ideas, I deliberately built up a diagrammatic summary as follows. (The positions are important.) Let the game continue for long enough to get the expected result so you can draw in the arrows correctly.

This stage is important, for it gives students a chance *at their own pace* to grapple with the results. Don't give hints about how this might be done; leave it to their initiative.

beats ⟶ equals ═══

Let's show the results. Firstly Red and Blue.

RED ⟶ BLUE

Then Blue beats Black and Black beats Green.

RED ⟶ BLUE
↓
GREEN ⟵ BLACK

This is the perfect place to pause, dramatically!

If Red is better than Blue, Blue is better than Black, and Black beats Green, then Red should absolutely thrash Green!

Surprise, surprise! Green beats Red! I put this new arrow on the diagram, to unbelieving looks from the class!

RED ⟶ BLUE
↑ ↓
GREEN ⟵ BLACK

We then added the two remaining arrows, to produce the complete 'answer'.

RED ⟶ BLUE
↑ ╳ ↓
GREEN ⟵ BLACK

It is such a surprise because it defies the expected transitive property: if A>B, and B>C, and C>D, then we expect A>D. Pupils are so used to this occurring with numbers, they expect it to occur always. (It does mostly, but not always, as this activity shows.)

4 Scissors, rocks and paper

While the students were pondering this improbable result, I had the opportunity to explain the 'non-transitive' (or circular) pattern by recalling this playground game.

I let students play the game in pairs for a few minutes, nostalgically.

Did you ever play a game about scissors, rocks and paper? Can someone remind us how it goes?

Scissors beats paper because scissors cut paper.

Paper beats rock, because paper covers rock.

Rock beats scissors, because rock bends scissors.

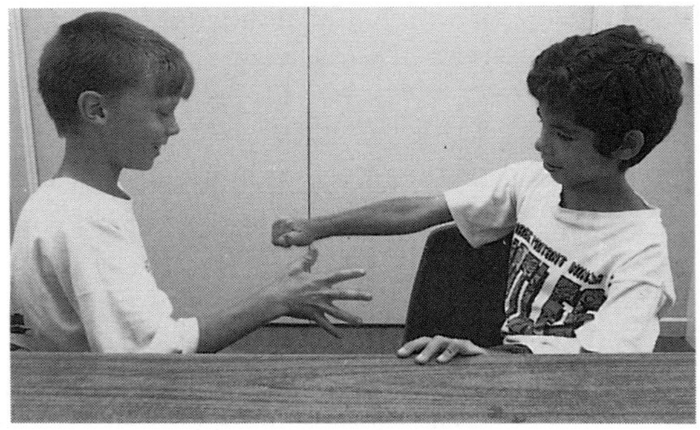

It seems that the game in some form is universal, and has continued for generations.

How are the dice like the scissors, rocks and paper game?

The parallel between the game and the dice is that they are both non-transitive; whichever item you choose, there is always another that can beat it.

Link this unexpected outcome to real applications; if they think of other examples, then they understand the idea.

Could the same happen with sports results?

It is relatively common in sports for team A to beat B, B to beat C and then for C to turn around and beat A. Can you find a recent, well-known example?

5 Problem solving

We have the results of our experiment. Try to work out why they occurred. For example, how could you explain why Red is better than Blue?

This section tries to move from the experimental stage to the more formal analysis stage.

I organised the students into groups again to discuss this.

It is like lots of different little competitions.

If Red rolls a 0, it loses every time.

If Red rolls a 1, it loses every time.

If Red rolls a 7, it wins four and draws two.

If Red rolls an 8, it wins all six times, so will the other 8.

If Red rolls a 9, it wins all six times.

So altogether, Red should win 22, lose 12 and draw 2.

It is a key question mathematically, since it aims to stimulate a discussion about methods of analysis. Don't suggest too early how to go about the analysis — see what happens. All can learn from seeing how others tackle the problem.

The key to the understanding is the intuitive phrase, 'it's like lots of different little competitions', which they can then systematically consider.

6 Theoretical analysis

There are several useful methods of analysis, the choice depending on the background of students.

The diagram helped make the type of explanation above clear to all.

		BLUE					
		5	5	6	6	7	7
RED	0	B	B	B	B	B	B
	1	B	B	B	B	B	B
	7	R	R	R	R	=	=
	8	R	R	R	R	R	R
	8	R	R	R	R	R	R
	9	R	R	R	R	R	R

You get the same result considering the Blue. If you roll a 5, you win two and lose four; so does the other 5. If you roll a 6 it also wins two and loses four; so also does the other 6. If you roll a 7 it will win two, lose three and draw one; so does the other 7. Adding these up gives 12 wins, 22 losses and two draws.

From this, it can be seen that Red is expected to win 22 out of 34 or 65% of the time.

Once we had shared this approach I set as tasks the analysis of the remaining five competitions between pairs of dice.

Tell me what happens if Black plays Green.

Normally, this sort of analysis is done at Year 11. But at that level it comes with all the jargon of sample spaces, equally likely outcomes, relative frequency, etc.

At higher levels, students might complete a symbolic table such as a 6 x 6 table showing all the possible outcomes of the rolls of the two dice. (This appears on the screen as part of the computer extension.)

7 Computer simulation

The computer can be used as both a reinforcement for the theoretical results above and as an investigative tool in the extensions. The program (if available) allows you to choose from the four special dice or to invent your own. It also allows you to run a certain number of rolls or compete up to a certain winning total.

Demonstration

Enter the numbers on the faces of the dice.

a Run a game on the computer to see which die wins.

b Run ten games and see which die wins most often.

Student activity

Students repeated their own dice rolling experiment using the simulation to play many games. This allowed them to check the theoretical frequencies they have calculated above.

I find it is wise to describe what the program does using the same terms as the activity just done.

It is important that pupils see the program as simulating exactly what they did with the dice.

8 The search for another die

Have you ever invented a new thing to add to the scissors, rocks and paper game?

I used this question to seed the thought of finding another die to extend the sequence. In the playground game, children have invented:

- bomb, which blows up rocks and paper, but scissors cut the wick;
- water, which rusts scissors, wears away rocks, but is absorbed by paper.

Can you invent new dice to try to fit into the dice game?

Students used both theoretical calculations and the computer simulation to provide evidence to test their choices.

9 Discussion and report

Since we had played many games and had a lot of fun, I felt it was important that the children realised how much they had learned. So I asked them to put it into their own words, and then write a brief report on the activity, and what they had learned, for their parents.

What are the mathematical things we have done in this investigation?

This included simple skills, like making tables and drawing graphs, and finding percentages. They also used more complex personal skills, such as group cooperation, the planning of an investigation, problem solving and operating the computer.

What mathematical ideas have we used?

Again the ideas were many, such as sampling, probability and relative frequency, and the surprise outcome — the idea of 'non-transitivity'. (See the *Content epilogue* for more discussion of this idea.)

10 Consolidation

This offers an attempt for children to look again at the same activity after the 'excitement' has worn off, and gain as much mathematical value as they can.

I handed out copies of the worksheet and children worked in pairs. After some time had elapsed, I discussed their answers, and then took up the written work for assessment purposes.

The worksheet shows two other sets of dice which are non-transitive. The children are taken through the analysis method step by step with the first set, and shown how to intepret the table. (This is the way the computer works it out.) Then they use the same method for the second set of dice and compare the two sets.

The second set of dice is very similar to the first! All we have done is rename the colours, and add three to each number. Since this makes no difference to the relationships in each 'play-off', the result can be easily predicted once the 'code' is cracked.

Once they have this idea, the answers to the last questions are obvious, and children can make up their own sets of dice with this property by simply applying any multiplication or division (except zero) and addition (or subtraction) to all the numbers on all dice.

11 Extensions

A: Rolling all four at once

The students were determined that one die 'had to be the best', and so devised for themselves a competition in which all four dice are thrown together and points allocated for the scores shown. This proved very interesting, as the outcome depended on how they defined the scoring system.

> For example, if you just score one point for the highest score, then Blue hardly ever scores.
>
> If you score four points for the highest score, three for the next, then two and one for the least, and fractions for ties, a rather different picture emerges.

B: Analysing the variability

This section arose from the observations that while one colour outscored another overall, it certainly didn't win all the time.

> I find that asking for estimates in advance of an experiment not only increases interest, but tells me about each student's understanding of what is going on.
>
> This issue is more complex than a straight competition between dice. It brings in the idea of chance and the variability of results.

Estimating

I wrote some of the wide-ranging estimates on the board.

How certain are we that Red will beat Blue?

Out of 100 rolls between Red and Blue, how often can we expect Red to win?

Experimenting

I asked each group to play one game (first to seven) between Red and Blue.

Red seems to be better, but Blue wins sometimes. What percentage of the time can we expect Blue to win?

They shared the results with the others by writing them on the board.

Computer simulation

We ran 100 rolls to discover the percentage of the time Red won. We did this again to observe the variation in results.

What percentage of the time can you expect Blue to have an 'upset win'?

> In a typical set of results Red wins most of the time. But in a few cases, Blue wins. Even though the scales are tipped against Blue, it is still possible for it to win! In sport this is called the upset win.

C: First to how many?

What percentage of games will Blue win, if the game is 'first to seven points'?

Is it the same as when the game was one roll long?

I wrote the guesses on the board.

Compare 'first to seven' with 'first to two' or other possibilities.

> This is clearly a new direction for the investigation. There is a computer option that enables you to see whether the game length affects the percentages of wins for each die.

Content epilogue: background information for teachers

Non-transitive

Transitivity is a fundamental property that is true of numbers: if A is bigger than B, and B is bigger than C, then we are sure that A is bigger than C. This is 'commonsense'!

It is only with something like dice, where there are six numbers involved on each, that someone clever can design a non-transitive situation.

Relationship arrows

We have used 'relationship arrows' to show this strange situation on a diagram. These arrows are often used in situations where we want to show a relationship between sets of things. Here is an example involving simple multiplication.

Start at any digit, multiply by 2 and if the answer is over 9 take the ones digit only. Here is the completed picture. Try it yourself for other multiplications.

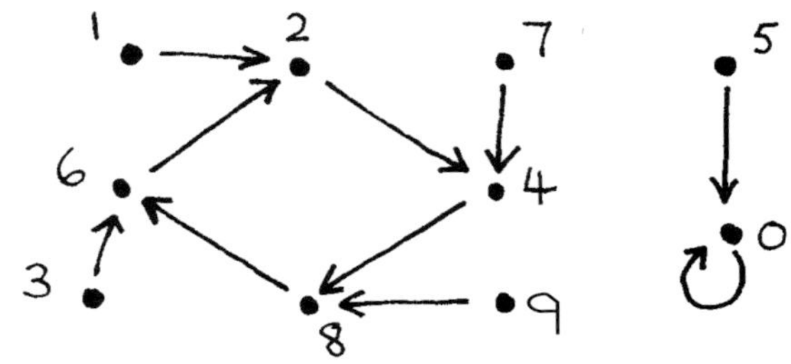

Probability and distributions

In the long run, after many rolls, one die can be expected to beat, or be equal to, another die. We can use a table to predict what fraction of the rolls will be won, and this table is given in the text for one game: Red vs Blue. However, for only a few rolls either die could win. It is this unpredictable feature (often called natural variability) that makes dice rolling so exciting.

If we play games of ten rolls each (Red vs Blue) and do it many times, Red will tend to win about six or seven out of each ten. But after many games it becomes clear that there is a spread of results. For example, Red will sometimes win three or four out of ten, and sometimes up to nine or ten out of ten.

It is the times when Blue wins that we have called 'upset wins'. About 35% of the total rolls will result in upset wins.

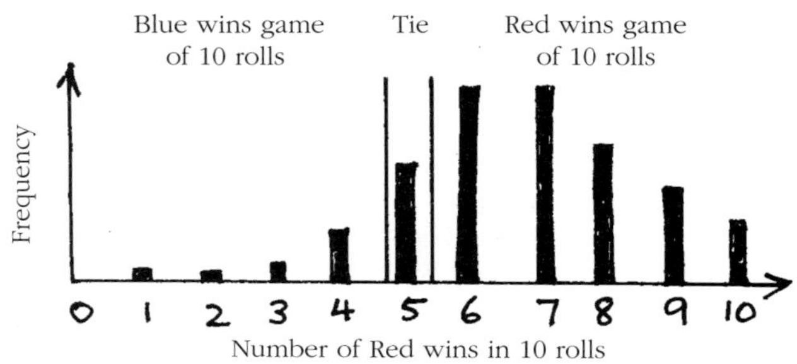

Effect of longer games: the binomial distribution

Red wins 65% of the time when only one roll needs to be won to win the game. So Blue will win about 35% of the time. If the game is 'first to two', then, at most, three rolls are needed (apart from ties, which don't count in the score). Red has an advantage in each roll and should win roughly two out of three (65%) separate rolls.

The tree diagram below is used to show the six possible outcomes of a game. There are three ways that Red can win. Probabilities are given and added to get the probability that Red wins a game of 'first to two'.

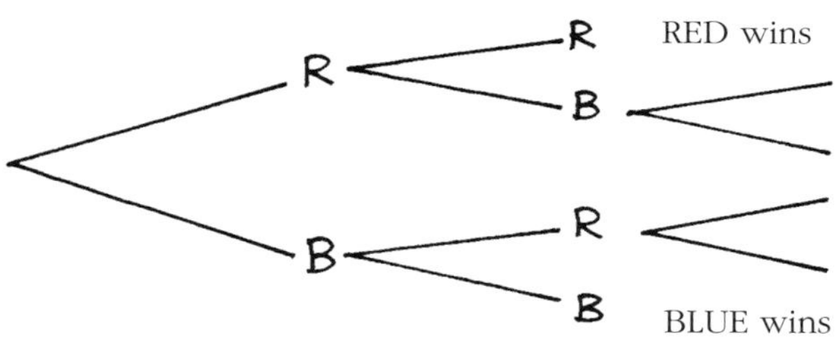

RED wins Pr = 0.65 × 0.65 = 0.423

RED wins Pr = 0.65 × 0.35 × 0.65 = 0.148

BLUE wins

RED wins Pr = 0.35 × 0.65 × 0.65 = 0.148

BLUE wins

Pr (Red wins 'first to 2') = 0.423 + 0.148 + 0.148 = 0.72

By making the game longer, we have given Red an even greater advantage. By the same logic there is even more advantage for Red to play longer games. By the time we get to first to seven, for which there may be 13 rolls, the chance for Blue to win is very greatly reduced.

Worksheet

Here is another set of four dice. Instead of rolling them, and finding the winners in that way, you are challenged to find the winners of each play-off by analysing the numbers on the dice.

Red: 0 0 4 4 4 4 Blue: 3 3 3 3 3 3 Green: 2 2 2 2 6 6 Black: 1 1 1 5 5 5

Let's do a play-off between Red and Blue. We take the first number on the Red die, and play it against each of the numbers on the Blue.

1 Complete this table to show the winner of each little play-off.

		Blue					
		3	3	3	3	3	3
Red	0	B	B				

2 Now repeat the same idea for each other number on the Red die. Take one line at a time.

		Blue					
		3	3	3	3	3	3
Red	0	B	B				
Red	0	B	B				
Red	4	R	R				
Red	4						
Red	4						
Red	4						

3 How many games will actually count in the result?
How many games are expected to be won by Red?
What fraction is this of the total games that count?
What fraction of games are expected to be won by Blue?
So now you know which of Red and Blue is the 'better'.

4 Use the same method to compare the others: Red vs Green, Red vs Black, Blue vs Green, Blue vs Black, Green vs Black.
Share the work with others.

5 Put all your results together and show how the dice compare on the diagram. (You may want to redraw the diagram with a different arrangement of colours.)

RED	BLUE
GREEN	BLACK

6 Use the same method to compare these four dice:

Red: 5 5 5 5 9 9 Blue: 4 4 4 8 8 8

Green: 3 3 7 7 7 7 Black: 6 6 6 6 6 6

RED	BLUE
GREEN	BLACK

7 Is there any similarity between the sets of dice in questions 5 and 6?

8 What happens to the relationship between the dice if you add one to each number on all four dice?
What happens to the relationship between the dice if you double each number on all four dice?

How long is a rod?

Years 7 to 12

Before the days of well-defined units of measure, an English unit called the 'rod' involved adding the foot lengths of 16 men. But foot lengths vary; how could this 'unit' be the same all over England? How much variation might be expected to occur? Just how long is (or was) a rod?

Features

Teaching and learning

- Historical significance of informal units
- Problem-solving strategy: simplifying the problem
- Group work
- Computer extension

Major mathematical content

- Comparison of two distributions (males versus females)
- Random sampling (die and random digits)
- Use of box-plots (median and interquartile range graphically)
- Average of a random sample is a better estimate than a single value
- The larger the sample the better the estimate

Mathematical content is further discussed in the *Content epilogue*.

Preparation and equipment

- Copies of the worksheets
- Each student also needs a centimetre ruler and/or a piece of graph paper
- Two packs of cards for stage 3

Expected timing

- From three to six sessions

A teacher's story

I came across the following interesting definition of a *rod* in an old book. It suggested opportunities to discuss old units, bring in some social history, and even alert students to some sexism (why only 'men'?). A rod was used for surveying land so it needed to be almost the same all over England.

> Take 16 men, short men and tall ones as they leave church and let each of them put one shoe after the other and the length thus obtained shall be a just and common measuring rod.
>
> (Jacob Köbel, *Geometrie*, 1536)

This definition shows that the mathematicians who advised the King were aware that the average of a random sample of foot lengths must be close to the actual average of all the foot lengths in England. If the average of a sample of 16 gets you close to the true average, then the sum of 16 random values will be close to 16 times the true average, and hence almost the same all over England.

I decided to ask the students to collect their own data in order to investigate the idea. While the definition uses the length of adult male feet, I wanted an investigation which would not involve the students in difficult data collection, so I modified the activity to use the handspans of the students, later extending this to foot lengths. On the way we checked on the idea that males and females have differently sized hands and feet, developed some ideas about random sampling, and in particular developed a fundamental concept in statistics: that the averages of random samples are less spread than the values from which they are taken, and that the larger the sample the less the spread and the better the average.

(In fact, the averages of random samples are 'normally distributed', with the true population mean as its mean. The spread depends inversely on the sample size. These ideas are explained in the *Content epilogue*.)

Comments

The curious start, about 16 men lined up outside a church, got them in right away! We took a good look at many of the old 'imperial units' while we were at it.

We had a great deal of fun measuring foot lengths with shoes and socks off. It surprised them to learn that the 'rod' must have been virtually the same all over England.

I liked the informal way the ideas leading to the Central Limit Theorem were introduced. Students were telling me about the 'normal' distribution of the means, and how they are far less spread for larger samples. For the first time ever, I think they really understood!

Summary of the stages in this investigation

1 **Introducing the problem**

2 **Offering a simpler problem: handspans**

3 **Introducing the concept of sampling**

4 **Sampling in groups**

5 **Reporting and discussing implications**

6 **Random sampling of the whole class**

7 **Using a computer to extend the investigation**

8 **Exploring foot-length values**

9 **Sampling 'rods'**

1 Introducing the problem

I started this lesson by discussing the units of measurement in the early days.

Who can tell me about some units of length measurement before the metric system?

Expected answers include:

foot (length of), inch (width of a man's thumb), yard (arm stretch), chain (22 yards), mile (1760 yards),...

The simplicity and convenience of using a body dimension as a measuring standard is easily recognised. Students may be interested in leading the discussion in other directions, such as capacity units (gallon, pint), weight units (ounce, pound) and so on. This is a good signal for a project; but before setting it and a deadline, have a word to your librarian.

I handed out Worksheet 1 and we discussed the illustration. I introduced and discussed the definition of a *rod.*

Here is one measure which was used in the 16th century. How long is it?

What do you think it might be used for?

Their ideas were intriguing, but frequently inaccurate. The rod was used to measure land, and for simple surveying.

At this point I organised 16 volunteers to stand and estimate for us the length of a rod. It was a bit noisy, but fun.

Why add 16 foot lengths? Why not just use one foot length and multiply by 16?

Students pointed out that if just one foot length is chosen it may be a particularly small or large one, but when many are added extreme values of length won't have as much impact.

Why just involve men? Is this being sexist or are their foot lengths different from women's foot lengths?

As a point of discussion, it is worthy of note that in later years a rod was also known as a *perch* or a *pole*, and defined to be 5.5 *yards*! This is equivalent to about 503 centimetres. At 22 yards, a cricket pitch is actually 4 rods long! Most students are aware that an imperial foot is about 30 centimetres , so 5.5 yards is 16.5 feet.

To follow this through we need between 50 and 100 different foot lengths. How can we get these?

Students suggested the need to collect some data.

To make it easier, I required each student to collect three or four values of *foot length*, using their family, friends and other students from school.

I deliberately set up the data gathering at this stage, because it takes time. Several days should be left so that all students have a chance to add their data to the file. At this stage we did not fuss about adult males only. It will be instructive to compare to see the differences (if any).

How will this data set be collected? How will we measure lengths? Does it make any difference which foot we use?

Should the subjects take their shoes off? Do we measure to the nearest cm, mm, etc?

How many measurements do we need? How will we choose who to measure?

The aim of this discussion is for the students to appreciate that any sources of variation that could be controlled should be controlled, while at the same time endeavouring to replicate the age-old method. It is important that the students are involved in these discussions.

2 Offering a simpler problem: handspans

While they were gathering the foot-length data, I continued the investigation by introducing a parallel problem for which data is easily measurable. This led to the need to measure their own handspan, which is the distance they can stretch between the tips of the thumb and little finger along a ruler.

Do girls and boys have differently sized handspans?

How can we display our data to see whether girls and boys have different handspans?

Girls	Stem	Boys
	22	0
1	21	2 8
3 2 0	20	3 4 6 9
9 7 5 3	19	0 1 2 2 8
9 6 5 2 2	18	9
5 4 2 0	17	

They suggested a back-to-back stem-and-leaf display, since we had recently done one.

The units digits for the girls' and boys' measurements are plotted on opposite sides of the same stem of tens digits. In this example, 20 I 3 means 203 mm.

The boys generally had hands that were much larger than the girls. The stem-and-leaf plot made it quite easy to find the median values for each (halfway in each set).

3 Introducing the concept of sampling

We had shown that males tended to have bigger hands than females, and possibly also bigger feet. Now the more subtle question had to be tackled.

> **Suppose you did not have all this data. You can only ask a small number of people to give you their handspans. How many people should you ask?**

Let's try choosing just one of you at random.

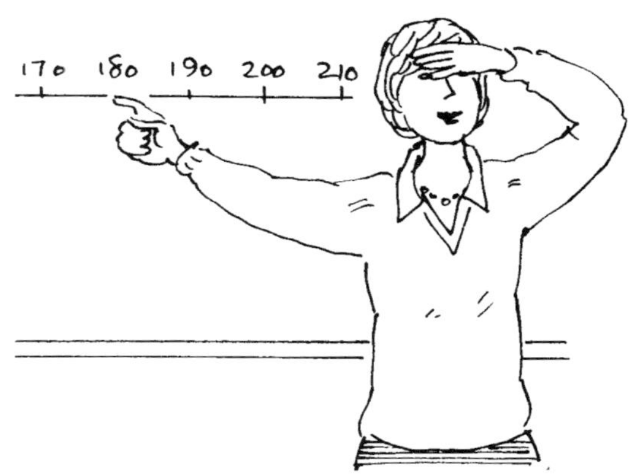

I 'covered my eyes' and turned around, then took a 'blind stab' at the board, choosing one of the very low (or very high) measures. This was not difficult, since I could cheat! I wanted students to see that I was cheating, so they would see the need for a 'fair' way to select even one person if it is to be 'at random'.

I then gave each person a card from a pack. I shuffled another pack, cut it and read out the card. It wasn't the same as any I had given out, so I shuffled and cut again until it was. The person with the matching card told me their handspan, which I wrote on the board.

I did this ten times, writing the handspans of each chosen person under the heading: *Estimates: samples of 1.* They were, of course, well spread out, and I graphed them using a simple line graph.

Samples of 1

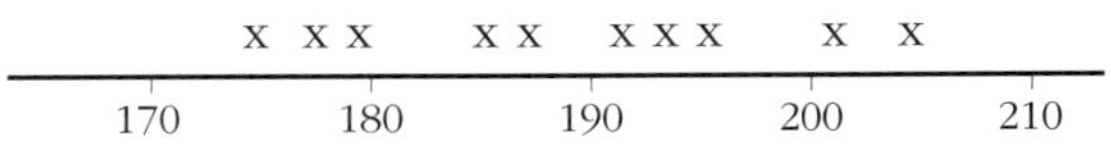

Now let's try choosing two and averaging.

I used the same method of choosing, but this time after I had chosen each pair, we averaged the values. I did this ten times and again graphed averages on the board. This time the values were much closer together.

Samples of 2

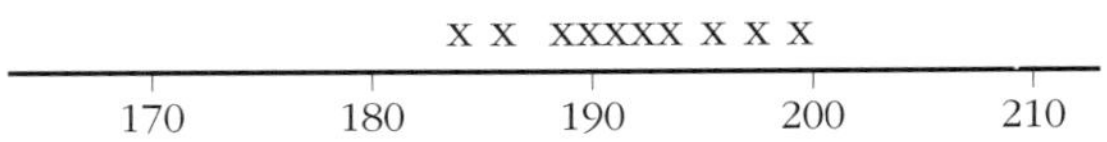

4 Sampling in groups

I handed out Worksheet 1 and went over the instructions. I set a deadline of about 20 minutes. Groups worked quickly, following the model provided by the previous classwork.

Work in groups of five or six, and roll a die to choose whose handspan you will use. (If there are five of you and you roll a six, ignore it.)

Students get a better idea of what is going on if they try it themselves. This will also allow them to discuss what is going on in groups, and to put ideas into words.

5 Reporting and discussing implications

When it came time for each group to report, they all had the same story to tell.

The averages were all very close to each other...and also very close to the true average.

The averages of four numbers were closer than the averages of two.

The explanations for this were in terms of the way the average balances out the larger or smaller values; it does this better if there are more values involved in the average.

6 Random sampling of the whole class

I now extended the findings to even larger samples using the whole class data. There were enough in the class to allow each person to choose a sample of ten. Learning how to do this was interesting.

Here is how you can choose your own individual random sample of ten handspans.

They copied each handspan in the class onto the table on Worksheet 2 so that each corresponded to a different two-digit number. They used random number tables to choose the two-digit numbers randomly and so obtain ten handspan samples, which they averaged.

You start at any number in the tables. Try to choose this randomly so you don't all start at the top left!

Take two digits in pairs, and if they give a number in our handspan table, write down the handspan in the second table on Worksheet 2.

Do this ten times, and take an average (add and shift the decimal point.)

When they had finished they told me the answers, and I wrote them all down.

They were amazed that they were almost all identical! However, don't be fazed if they are not; they still should be very close.

We clearly have agreement on an estimate of the average handspan of the class. Now what is the true value?

I used this easy way.

- Each group adds together their own lengths, and we put them on the board.
- We all add the group totals on the board, then use a calculator to divide.

The value is no surprise, as it has been predicted using the sampling methods above.

7 Using a computer to extend the investigation

You need to enter the set of numbers from which the computer will draw its sample, or ask it to choose its own set of numbers.

The computer program is able to choose random samples of any size from any set of values, and find their averages (means). The program can do this many times and graph the results. The means are shown to be clustered around their own mean value, which is extremely close to the 'true mean' of the values being sampled.

Demonstration

I asked it to choose a random sample of two, and ran just one trial. It showed the two values it chose and displayed their average (mean). It also calculated the mean of all sample means; at this stage there was only one value, so it displayed that. It calculated the 'true population mean' for us; we saw that the average of one sample of two values was not very good as a prediction of the true mean.

I ran this 50 times. Each time it showed the two selected values on the graph, and their mean. It also showed a graph of the means themselves. We made a note of the spread of these average values. It also found the average of the means.

I changed the sample size to five. We ran 50 trials, and compared the spread of the means with the earlier values.

I changed to a sample of ten, ran 50 trials, and we saw the graph of the spread of the means.

Three points were worth noting; all are part of the Central Limit Theorem:

- The spread of the averages was much smaller with larger samples.
- The shapes of the graphs should all be quite close to a *normal distribution* curve. This was the case, no matter what the actual numbers in the original table.
- The mean of the means was very close to the mean of all the values in the population from which samples were being taken.

8 Exploring foot-length values

By this stage students had gathered enough foot-length data for us to use it in a similar investigation. As the students read out their lengths we recorded these on Worksheet 3, with the tag M or F to show male or female. Students were asked to write numbers into any randomly chosen vacant cell on the worksheet, so most, if not all, sheets were different.

Students first graphed the numbers using F or M so we could compare the two distributions.

Next someone found the true average, using a calculator. (The computer gave the same answer.)

I also entered the numbers into the same computer program as data. (In order to compare male foot lengths with female foot lengths, they have to be treated separately.) This allowed groups of students to continue the investigation using the computer instead of the longer method.

9 Sampling 'rods'

We know what the true average foot length is; the true 'rod' length should be 16 times that. Now we have to add samples of 16 foot lengths and see how close they come to the true 'rod' length.

All students used the random number tables (from Worksheet 2) to select their own 16 foot lengths. They added their own 16 numbers to give their own estimate of the length of a rod.

Who has the smallest? Anyone under...? Who has the largest?

Tell me your estimates and I will graph them on the board.

I used a line graph to display the results. As expected, there was little variation, showing that the old unit of rod was likely to vary only slightly around the country.

Content epilogue: background information for teachers

Random sampling

The idea of 'random' is not an easy one. It seems to be related to the idea of fairness; if all possible values have an equal chance of being selected, then the process that does it can be called random. We see one method on TV when Lotto and Keno numbers are selected.

Students are frequently surprised when they take a small random sample by a method that seems fair (such as by rolling a die or tossing a coin) and find that the numbers are not selected evenly. This is a common situation, called natural variability. It is only in the very long run that the results even out.

Methods can include gambling tools or random number tables; the last four digits of phone numbers are commonly available. Some calculators provide random decimals between 0.000 and 1.000.

Comparing two distributions

The spread of male and female handspans probably overlap by quite a bit. How can we tell whether the sets of numbers are really very different? If you have graphed all the values there are, you can just compare the averages. But even then, if they overlap a lot, how can you be sure?

However, if the values are random samples you may not have taken very representative samples. The usual way to deal with this is to calculate the means (averages) of the two sets and to compare the difference between them to the 'average spread' of each of the groups. There is a complex formula that tells whether the sets are likely to be really different or whether any observed difference is just a result of the sampling.

The Central Limit Theorem

One of the most important ideas in statistics is addressed in this investigation. It says that, no matter what the spread of numbers, the averages (means) of random samples are normally distributed, and the mean of that normal distribution is the same as the mean of the original population.

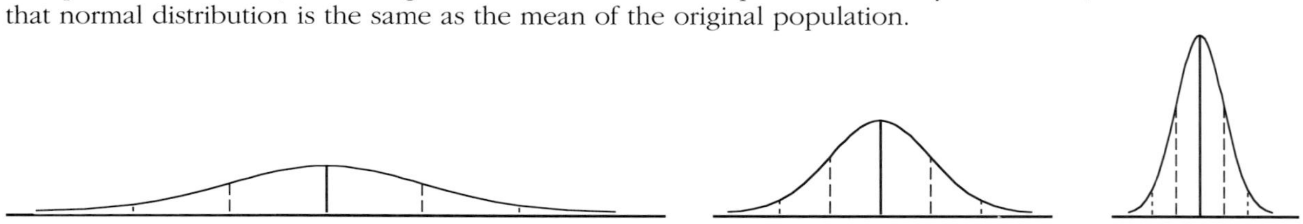

Sample of 1 Samples of 4 Samples of 6

In addition, it says that the larger the random sample, the smaller is the spread of the sample means. The standard deviation is inversely proportional to the square root of the sample size used. So you find the standard deviation for samples of one; then the spread for samples of four is half of that, and for samples of 16 is half again.

Here is some typical data, in case you need it. You could compare your class to this one.

Sample foot-length data: mean = 31.5 cm

30.3	30.9	31.2	32.3	31.3
30.7	32.8	31.0	33.3	30.7
32.2	30.1	31.6	32.1	31.4
31.8	32.9	31.9	29.4	31.6
32.1	31.2	30.7	32.1	30.8

Sample rod data (in cm):

499.7	504.5	506.7	504.6	502.6
504.9	503.9	501.9	505.0	502.3
501.6	502.0	502.5	499.7	504.0
495.1	500.1	497.4	503.0	499.4
500.8	499.5	497.5	498.4	506.4

Stem-and-leaf plot of sample rod data:

495	1
496	
497	4 4
498	3
499	4 4 6 7
500	7
501	6 8
502	0 2 4 5 9
503	9 9
504	4 5 9
505	0
506	3 6

mean = 501.7 cm

Worksheet 1

> Take 16 men, short men and tall ones as they leave church and let each of them put one shoe after the other and the length thus obtained shall be a just and common measuring rod.
> (Jacob Köbel, *Geometrie*, 1536)

1 Graph all your handspan values using a line graph.
All handspan values:

2 Choose two using the die, and average them. Do this ten times and graph the results below.

	Choose 2	Average
1		
2		
3		
4		
5		
6		
7		
8		
9		
10		

3 Choose four using the die, and average them. Do this ten times and graph the results below.

	Choose 4	Average
1		
2		
3		
4		
5		
6		
7		
8		
9		
10		

Samples of 2: line graph

Samples of 4: line graph

4 Find the average of all handspans in your group:

5 When you have finished, be ready to report what you have discovered.

Worksheet 2

01	02	03	04	05	06	07
08	09	10	11	12	13	14
15	16	17	18	19	20	21
22	23	24	25	26	27	28

	Choose ten										Average
1											
2											
3											
4											
5											
6											
7											
8											
9											
10											

Random numbers

15	**19**	**11**	**4**	**21**	**7**	**20**	**16**	**2**	**23**
22	**25**	**26**	**10**	**24**	**1**	**8**	**6**	**30**	**9**
3	**28**	**13**	**14**	**29**	**5**	**12**	**27**	**17**	**18**

Worksheet 3

00	10	20	30	40	50	60	70	80	90
01	11	21	31	41	51	61	71	81	91
02	21	22	32	42	52	62	72	82	92
03	13	23	33	43	53	63	73	83	93
04	14	24	34	44	54	64	74	84	94
05	15	25	35	45	55	65	75	85	95
06	16	26	36	46	56	66	76	86	96
07	17	27	37	47	57	67	77	87	97
08	18	28	38	48	58	68	78	88	98
09	19	29	39	49	59	69	79	89	99

Colas

Years 9 to 12

Can the students in your class really tell the difference between colas such as Coke™ and Pepsi™? How can you use statistics to tell whether they are guessing? Students get some insight into product testing and an introduction to significance testing.

Features

Teaching and learning

- Uses realistic context
- Uses firsthand data
- Demonstrates methods used in industry and commerce
- Tackles social issues of advertising
- Group work

Major mathematics content

- Graphs of frequencies as histograms
- Simulation
- Compares outcomes with chance (significance testing)
- Uses binomial probability distribution

Mathematical content is further discussed in the *Content epilogue*.

Equipment and preparation

- Two cans (or one large bottle) each of two different colas, or more if you do the three drink extension; keep them in the staff fridge until ready to do the taste test, and pour them out into the jugs just before bringing them into class
- Two (or three) jugs with labels A, B (and C), and a basin with cold water for rinsing
- Plastic or paper cups, one per student.
- Copies of the worksheets
- Coins for stage 5, although students could be expected to supply these
- At least one computer; making sure you have tested the program
- Dice for extension E

Expected timing

- Two to five sessions

A teacher's story

This sequence of lessons develops some advanced concepts and shows how probability links to statistics to help people make decisions on the basis of a small amount of data. It also links to the commercial world through the process of advertising. The two rival brands of cola are in a worldwide competition to attract sales. They both advertise extensively, and hope to establish brand loyalty. One of the outcomes of the investigation may be to discover the impact of advertising.

I first asked students to tell me which of the two rival colas they preferred. Then they came out and tasted a little of each from jugs, and wrote their decision about which was Coke™ on the board. At this point I didn't tell them the answer.

I suggested that some of them may be guessing. To compare their percentage correct with pure chance, they guessed whether a coin will fall heads or tails. I tossed and we recorded the percentage who were correct. We did this a few times and got a range of results. We found that quite high percentages of correct answers are likely by pure guesswork!

I announced the correct answer, which led to the class deciding whether or not their result could be expected to happen by chance. We saw whether we could tell the difference between the two rival colas. If not, any preferences that we may have for one over the other may be purely due to advertising images!

In the extensions, we used a computer simulation to expand this idea and show the distribution of the percentage of correct guesses. We looked more closely at how the number of people in the group changed the distribution of the percentage of correct answers. Students realised that the spread of 'percentage correct' is narrower with larger groups. For a group of five, they need *all* to be correct to be reasonably sure that they did not guess. For a group of 20, only 70% correct is sufficient for most people.

The extension to three colas represented choosing between six possible orders, and hence could be simulated with dice. The taste test for three colas showed whether the students could really tell the difference between all three!

Comments

The students enjoyed the novelty of the tasting; the coin tossing contrasted well. It was a good chance to develop a practical approach to maths, a subject too often regarded as dry and theoretical.

I used this investigation on a maths camp with Year 8 to Year 10 students who were fascinated. It encourages me to a less formal approach to chance, drawing on the students' intuition.

Summary of the stages in this investigation

1 **Introducing the problem**

2 **Can you tell the difference?**

3 **One-person blind test**

4 **Whole-class blind test**

5 **What happens if you just guess?**

6 **What percentages are likely by guessing?**

7 **Interpreting the results**

8 **Are some of us guessing?**

9 **Discussion and report: what have we learned?**

10 **Consolidation**

11 **Extensions**

1 Introducing the problem

I discussed the world-wide competition between the two colas, and the importance of advertising.

You all know the two rival cola drinks: Coke™ and Pepsi™. Which do you prefer?

I wrote these straw-vote figures on the board so I could refer to them later and handed out Worksheet 1 for the students to record them also.

Can you recall any of the advertisements for either drink?

This introduction provides evidence of the popularity of the two drinks. The rest of the lesson will help the students determine whether they can tell the difference or whether the popularity is based on an advertising image.

One ad for Pepsi (in 1990) actually implied that you can taste the difference. The famous blind singer/pianist Ray Charles is playing. During a pause he reaches for a can. After a taste he realises it's Coke and not Pepsi. His response: 'Who's the wise guy?'

2 Can you tell the difference?

Can you *really* tell the difference, or are you just guessing?

I offered them a chance to investigate this question with a method used by the two companies themselves: blind testing. As we did the testing we also used ideas from probability and statistics. There are four stages involved; I wrote them on the board.

This introduction to the lesson is designed to increase their sense of participation by letting them know what is ahead. The novelty of cola tasting certainly increases anticipation.

One-person blind taste
Whole-class blind tasting
What happens if you guess?
How many correct to be confident?

3 One-person blind test

We then simulated the advertisement. I chose one person to do a 'blind' test. The two colas were available in separate jugs, clearly labelled A and B. I asked the person to say which jug contained the Coke, and wrote the answer on the board.

If one person gets it right, does that prove that the person is not guessing? Why or why not?

The point here is that one person might be guessing, so whether they are right or wrong doesn't tell you whether they can really tell the difference. It proves nothing.

4 Whole-class blind test

How might we tell whether people in general can tell the difference?

I had arranged for enough plastic cups for one per person. For a class of 20 two cans of each drink seems to be just enough. They took a small taste of each, and decided for themselves. I gave out Worksheet 1 for students to use.

When you have decided, take a slip of paper, write A or B for the jug you think contains Coke, and put your vote in the box.

They voted secretly, and the summary went onto the board. I worked out the percentage correct, without comment.

> It seems important that each person gets to vote without influence from others. This takes a while, and there is much quiet discussion as they decide.

5 What happens if you just guess?

Let's suppose you were all guessing. How can we find out what results you might have obtained?

I will toss a coin and I want you to decide in advance whether it will show head or tail. Write down your guess now.

How many of you do you expect will have guessed correctly?

I tossed the coin, and kept it hidden.

> Most students guessed half the class: 50%.

I told them the outcome. We counted the number correct, converted it into a percentage with a calculator and wrote the percentage on the board.

6 What percentages are likely by guessing?

Why did you expect about 50% of each group to be correct? If we run this many times will 50% always be correct?

I tossed again, and told them the result. Again the percentage went on the board and I organised these into a simple stem-and-leaf plot using stems of the tens digits (2, 3, 4, 5, 6, 7) and leaves of the ones digits. Students copied this onto Worksheet 1.

We repeated the activity until we had 20 percentages on the board and on Worksheet 1.

> Running this 20 times may seem like a lot of work, however it can go smoothly once you have routine, and it is needed to develop the idea that the percentages are *widespread around the 50% mark.*

7 Interpreting the results

Let's see what the guessing results tell us about our own taste test result.

How good does the result need to be before you are convinced it is not just luck? Is it 50%?

No, since this is the most likely result from guessing.

Is it less than 50%?

No, since most have the answer wrong; they can tell the difference, but it is the wrong way around!

Between 50% and 65%?

Guessing is most likely, since results up to about 70% are quite likely by guessing.

Or 100%?

Yes, if the sample is large; however, for two people, one in four results will be 100% by guessing.

So if 100% is Yes, and 65% is No, where is the cut-off? Can we say how good we have been? How tough are we going to be on ourselves?

8 Are some of us guessing?

Finally I told them which of the colas was Coke.

Could your percentage correct have occurred by chance? Does this show that some of you were guessing?

It is possible that your class can tell the difference, but not very likely. Many were probably guessing, and the percentage correct will almost certainly show that the outcome is no better than guessing. (In some classes, the percentage was actually less than 50%.)

Who was guessing?

One class picked the difference with 100% accuracy, which meant that it was necessary to modify the discussion! The teacher also did the taste test, and honestly selected incorrectly, so this generated much discussion.

I finished by indicating the vote they took earlier, in which they committed themselves to a decision that one was better than the other.

If you cannot really tell the difference between the drinks, why did you say one is better than the other?

If people can't really tell the difference, why is one drink more popular than the other? Are *you* a victim of advertising hype?

They were somewhat sobered to realise that the drinks are nearly identical, and their opinions were based only on advertising images!

9 Discussion and report: what have we learned?

After all this tasting and tossing, it seemed a good idea to make sure that students realised that they had been doing some very important mathematics. After we had discussed the ideas, I asked them to write a brief paragraph to describe 'significance testing'.

> Significance testing is the standard name for the total activity we were engaged in. We were trying to decide just how likely it was that a sample of results could have occurred by chance. (See the *Content epilogue.*)

10 Consolidation

This extra practice (on Worksheet 2) was set for homework several weeks afterwards; it served to review the major ideas by recalling them to students' minds as they worked through the process once more.

They had to think hard about the table, using the data on the graph. Their answers to the problems gave me a good idea of how much they had understood. In particular, eight right answers out of 10 is at about the 'critical' level, being about 5% likely to have occurred by chance.

The last problem recalls that, for guessing, the chance of all being correct is 0.1^{10}. This is not very likely, so if they all get it right you have excellent evidence that at least some of them were not guessing.

The last part of Worksheet 2 is for extensions C and D.

11 Extensions

A: Test another class

The class were interested in acting as testers, and gathering results from other classes. I sought the cooperation of several teachers in the school, and had small groups of students visit these classes with two jugs of cola and plastic cups. The students explained the purpose of the taste test to the class concerned, and used the same secret voting system to gather the results. After they had counted the slips, they announced the percentage correct, and explained to the class whether or not the result was likely to have occurred by chance.

> This additional responsibility served to consolidate the rather difficult concepts and provide practice in experiment management and the interpretation of results.

B: Computer simulation of guessing

Demonstration

I used the number in the class for the number in each trial, and ran one trial. When it was done, I showed them the graph, and we did several more, one at a time. The graph recorded the result for each trial and the distribution grew as the number of trials increased.

> The computer is very useful in providing much more data based on guessing, doing it quickly and graphing the result. If you do not have access to the computer program, students could use the table of random Hs and Ts (on Worksheet 1) and stem-and-leaf plots for their results.

Then I asked the program to run many trials, say 100. The graph shows the distribution of percentages correct if we are just guessing. (For a class of 20 most of the results are contained between 30% and 70%.)

Now we have a more complete picture of what happens by chance. Where should we put the cut-off?

I let them offer suggestions, with reasons.

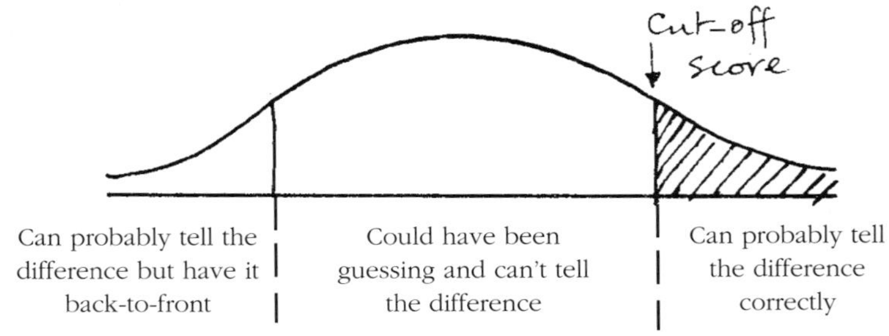

Someone pointed out that the cut-off will depend on how important it is to be right — reflecting the penalty if you're wrong. This illustrates a very important decision that must be faced by doctors every time they recommend a procedure for treating a patient. There are two ways to be right, and two ways to be wrong.

- If the doctor correctly diagnoses the disease, and the patient has the disease, there is no problem.
- If the doctor fails to diagnose the disease, and the patient does not have the disease, there is no problem.
- If the doctor fails to diagnose the disease, thinking that the symptoms may have occurred by chance (that is, some other cause), then the patient may suffer through lack of proper treatment.
- If the doctor diagnoses a particular disease where it does not exist, and the treatment has undesirable side effects, then the patient may suffer.

Mathematicians usually agree that, for most situations, if an event is expected to happen only once in each 20 times (that is 5% of the time) then it is unlikely to happen by chance.

What 'percentage correct' has only 5% of the results higher than it?

For my class size it was about 70% correct. For yours it might be different.

We set the cut-off score at this value and ran 100 trials again. It told us the percentage of the trials in which the percentage correct was better than, or equal to, the cut-off value.

C: Test a group of five

We have seen that if one person does the taste test we can't tell whether or not they are guessing. But if a whole class does the test, there is a cut-off, and a result better than that is unlikely to happen by guessing. This activity explores the situation for one smaller size of group.

What will happen if only five people take the taste test?

There are three options for this: computer, coins or random Hs and Ts. Because they could not all get to the computer at once, I started all groups on the coin method, and arranged for each group in turn to come out and use the computer.

Coin-tossing simulation

I gave the groups the following instructions:

- You each write a guess for the coin outcome (H/T).
- One person tosses a coin; all mark your guesses (√/x).
- You each record the number in the group who got the correct answer.
- After 20 such tosses, count the number of each and record in a frequency table.

They settled down to work on this task with no fuss at all. It seems I had given very clear directions and it led to an easily managed period of group work.

I handed out Worksheet 2. Each group carried out the same experiment until they had done it 20 times. I kept a watch on the groups, and when most had completed the work, I called a stop to the group activity.

An alternative method: work individually using random Hs and Ts

Worksheet 1 provides a large number of Hs and Ts, generated randomly on a computer. Instead of tossing coins, we can start at any (randomly chosen) point in the table and read to the right as normal. So to represent a group of five guessing the coin toss, we can look at five entries in the table, and count the number of Hs. This is the number 'correct', since H is the right answer! Do this 20 times.

I asked each group to tabulate its results on the board. For example:

Number	0	1	2	3	4	5	Group
Frequency	2	4	6	5	3	1	a
	1	2	6	7	2	2	b

We totalled the frequencies for each outcome, and drew a histogram.

Our results for 80 trials (20 from each of four groups) were:

Number	0	1	2	3	4	5
Percentage correct	0	20	40	60	80	100
Frequency	3	10	26	25	12	3
Percentage of total trials	4	13	33	31	15	4

The results will be similar, but differ slightly for each group. Help them to see that:
- for most probability situations, the 'expected' result will be the most likely outcome
- other outcomes are not only likely to occur, but will occur a predicable percentage of the number of times you run the experiment.

These are the basic ideas of a probability distribution.

What percentage of the time could four out of five get it correct by chance?

Is it reasonably likely that four out of five could get it right by chance?

What is the chance that *at least* four out of a group of five can get it right just by guessing?

What about five out of five correct? Is that likely to happen by chance?

By converting the totals to percentages of the total number of trials, we concluded that three out of five will occur about 31% of the time, four out of five about 15% of the time and five out of five only 4% of the time. So five out of five is quite unlikely to happen by chance. If you get all correct, we can say that you probably can tell the difference! Notice that we still can't be absolutely certain!

Computer simulation of five guessers

The computer program on the disk enables you to repeat this activity very quickly.
- As you are simulating coin tosses, choose probability of 0.5.
- Choose five for each trial and a cut-off percentage of 80% correct (that is, four or five out of five).

The program will run many trials and graph the results. It will also tell you the percentage of trials that have a value equal to or greater than the cut-off percentage.

With samples of five, there is quite a spread of the possible outcomes. Roughly 20% of the time *either* four *or* five out of five will be correct, but only about 3% of the time all five will be correct! The next step aims to show that these results depend on the size of the group.

Analysis (optional)

The expected results for five people guessing are fairly easy to analyse. The following steps are suggested.
- Systematically list all the possible combinations of results; a tree diagram may assist this.
- Collect together those that have five correct (1), four correct (5), three correct (10), two correct (10), one correct (5) and zero correct (1). These numbers are the binomial coefficients, found in Pascal's triangle.
- Because there are 32 in all, then one out of 32 (3%) will have five correct, five out of 32 (16%) will have four correct and so on.
- So 19% of the time you can expect either four or five to be correct by chance. This is quite likely, so that you must insist on five correct out of five for a group of five people.

D: Vary the size of the group

How big do you think the group needs to be so that you can tell whether they are guessing or not?

We found that one person's guesses are very widely spread (either 0% or 100% right), and 20 people's guesses are not spread far at all: the 95% cut-off is about 70% correct. This investigation shows that, the larger the group, the narrower the spread, and the harder it is to get 70% right by just guessing.

Here are four results from taste tests with different sizes of groups:

one person, 100% correct

five people, 65% correct

20 people, 80% correct

80 people, 60% correct

Which is most convincing?

It was clear that they understood the general idea: that larger groups had smaller spreads, and the fact that this made it less likely for the score to be over a cut-off. They saw the need to get more information to be sure about the 'most convincing' of these four situations. At this point the power of the computer became very much in demand.

I set them in groups to discuss this problem; it required a lot of thinking, and they needed to be able to discuss it with others. Then I asked one person from each group to report the group's thinking to the class.

Computer simulation

I ran 100 trials with the cut-off at 70%, and different group sizes. They recorded each result in the tables at the end of Worksheet 2 and I summarised the information on the board.

- group size of 5 — about 20% of the results will be on or over the cut-off
- group size of 10 — students will see a narrower distribution than before. The percentage on or over the cut-off is about 17%
- group size of 20 — the percentage on or over the cut-off is about 6%
- group size of 30 — the percentage on or over the cut-off is about 3%.

The computer program gives the results very quickly, and also draws the graphs for you. The program can be set to any group size.

The board summary looks like the table below, to highlight the differences. The cut-off is shown as a dashed line (70% in this case).

Summary of results

	Groups of 5		Groups of 10		Groups of 20		Groups of 30	
	% correct	Freq.	% correct	Freq.	% correct	Freq.	% correct	Freq.
	80–89		80–89		80–89		80–89	
	70–79		70–79		70–79		70–79	
	60–69		60–69		60–69		60–69	
	50–59		50–59		50–59		50–59	
	40–49		40–49		40–49		40–49	
	30–39		30–39		30–39		30–39	
	20–29		20–29		20–29		20–29	
Results (%) on or over cut-off:								

How do these distributions depend on the size of the group?

The spread is much less for larger samples. (In fact it is the spread for groups of one, divided by $\sqrt{n}$ where n is the sample size.)

For each sample size, what is the cut-off score we should use if we wished to have only 5% of values on or above it?

This question requires reading the tables carefully. To get more accurate values, you can go back to the computer. Try different cut-off values for particular sample sizes until you find the one that seems to give about 5% of the trials over the cut-off. (Note that this will change as it too is subject to random variation.)

E: Guessing from three choices

Extending the previous ideas to choosing between three colas is straightforward and provides reinforcement for students of the complex ideas involved. Instead of coins (two choices, equally likely) we can use dice, since for three colas there are six possible different orders (ABC, ACB, BAC, BCA, CAB, CBA) and only one will be correct.

The problem

In how many different orders could you match the drinks to the jugs?

I named the three colas to be matched to the three jugs A, B and C. (We used Diet Coke, Diet Pepsi and regular Coke.) The questions are designed to find out how much they picked up from previous experiences.

How many correct answers can we expect purely by guessing?

You will be correct 1 in 6 times, in the long run. This is about 17%.

Will we always get about 17% correct by guessing?

No, there will be a range of results, with 17% the most likely and most common.

What range can we expect? Does it depend on the number in the sample?

It surely does depend on the number in the sample, as the previous investigation will have shown.

How can we find out? What method could we use to simulate choosing between six equally likely events?

This lead to the design of a simulation. You could use dice, a sheet of dice outcomes printed by a computer, a regular set of random digits (ignoring 7, 8, 9, 0), or a computer program.

Computer simulation

We used the computer program described previously. To get one chance in six, we used 0.16666 for the probability, chose a group size of six, and ran 100 trials.

The point was to see how likely a group of people are to get a percentage correct by chance.

Let's increase the group size to see whether the spread of the distribution is smaller.

For example, if you go from samples of six to 24 (multiply by 4) the spread will be divided by two.

If there are six people guessing, what is the likelihood that these numbers will be correct by chance?

- **at least one (66%)**
- **at least two (26%)**
- **at least three (6%)**
- **at least four (1%)**
- **at least five (0.07%)**
- **all six (0.002%)**

The theoretical answers are given in parenthesis. They may or may not be close to your answers, depending on the number of trials you were able to run.

Now we simply repeated the tasting procedure used before. Students were again asked to write their decisions on the board. I wrote the correct match.

Together we went over the process of selecting *groups of six* students and deciding whether that group of six could tell the difference between the three colas. If we require that only 5% of the time the result could not be expected to occur by chance, then we needed to have at least three out of six correct.

Using our result, can our class tell the difference between the three colas, or are we just guessing?

For the *whole class* you can use the table below to decide the number of correct answers needed so that only 5% of the time the result could not be expected to occur by chance.

Class size	12	16	20	24
Number correct	≥5	≥6	≥7	≥8

Content epilogue: background information for teachers

The binomial distribution

Whether there are two or three colas, there are two choices. The matching you make (guess or otherwise) is either right or wrong. Let's look at the two colas situation first. Suppose that the chance of a correct guess is 0.5. Then for one person guessing, the probability of 0 correct is 0.5 and the probability of 1 correct is 0.5. This is shown in the first line of the table below. The sample size is n.

If two people are guessing, there can be none, one or two correct. There are two separate ways of getting one correct, depending on which person gets it right. So the probabilities are 0.25 for none and two correct, and 0.5 for one correct (the second line of the table).

The first table is easy to calculate, using powers of 0.5 and the numbers from Pascal's triangle, shown in bold type on the table below.

	Number correct				
n	0	1	2	3	4
1	0.5	0.5	–	–	–
2	0.25	0.5	0.25	–	–
3	0.125	0.375	0.375	0.125	–
4	0.0625	0.25	0.375	0.25	0.0625

	Number correct				
n	0	1	2	3	4
1	**1**(0.5)	**1**(0.5)	–	–	–
2	**1**$(0.5)^2$	**2**(0.5)(0.5)	**1**$(0.5)^2$	–	–
3	**1**$(0.5)^3$	**3**$(0.5)^2(0.5)$	**3**$(0.5)(0.5)^2$	**1**$(0.5)^3$	–
4	**1**$(0.5)^4$	**4**$(0.5)^3(0.5)$	**6**$(0.5)^2(0.5)^2$	**4**$(0.5)(0.5)^3$	**1**$(0.5)^4$

For the three colas, the probability of being correct is one in six (or 0.166), and of being wrong is five in six (0.833). So the table is as follows.

	Number correct				
n	0	1	2	3	4
1	**1**(0.833)	**1**(0.166)	–	–	–
2	**1**$(0.833)^2$	**2**(0.833)(0.166)	**1**$(0.833)^2$	–	–
3	**1**$(0.833)^3$	**3**$(0.833)^2(0.166)$	**3**$(0.833)(0.166)^2$	**1**$(0.5)^3$	–
4	**1**$(0.833)^4$	**4**$(0.833)^3(0.166)$	**6**$(0.833)^2(0.166)^2$	**4**$(0.833)(0.166)^3$	**1**$(0.166)^4$

If you recall the algebra of binomial expansions, you will see that the lines in these tables are the terms of the expansions of the binomial expressions shown at right. The values of q and p are the probabilities of a wrong and a correct guess. Hence p and q add to 1.

This relationship to the binomial expansion is the reason it is called the binomial distribution.

n	Expression
1	$(q + p)^1 = q + p$
2	$(q + p)^2 = q^2 + 2qp + p^2$
3	$(q + p)^3 = q^3 + 3q^2p + 3qp^2 + p^3$
4	$(q + p)^4 = q^4 + 4q^3p + 6q^2p^2 + 4qp^3 + p^4$

Spread of the binomial distribution

We find that the spread of the distribution gets smaller when larger sample sizes are involved. The spread is often expressed in terms of the *standard deviation*; about two thirds of the values will lie one standard deviation either side of the mean. The size of the standard deviation depends on the inverse of the square root of the sample size, so the spread is halved if the sample size is four times as large, and so on.

Significance testing

We spent some time in the investigation developing the idea that a result could have occurred by chance (guessing). To decide if this was likely, we have to know what percentage of the time, in the long run, a particular result would happen if guessing was involved. The usual convention is to choose a result value large enough that it would only happen by chance a small fraction of the time. If that result then occurs we think it may not have been by chance. For example, with 5% as our fraction, that result will only happen 5% of the time by chance (or one in 20). This is fairly rare, so if it happens, we say that it was not by chance. (Since it might have been; making that decision condemns us to be wrong about 5% of the time!)

Worksheet 1

Number in class ☐ Preference: Coke ☐ Pepsi ☐

Tasting

My answer: Coke is in jug ☐ Result: Coke is in jug ☐

Percentage in class correct:

Interpretation:

Guessing (20 coin tosses)

Class guesses (% correct):

___ ___ ___ ___ ___

___ ___ ___ ___ ___

___ ___ ___ ___ ___

___ ___ ___ ___ ___

Stem	Leaf
7	
6	
5	
4	
3	
2	

T T H T H H T H H H T T H T T T H H T H H T H H T
H H T H H T H T H T T H H T T T T H T T T H H T H
H T T H T H T H T T T T H H H T T T H T T T H T H
H T T H H H H H H T H H T T T T T T H T H H H T T
T T T T H T T H T H T H H T T H T T T T H T T T T
T T T T H T T H T H T H T T H H T H H H H H T T T
T H T H H H T H T H H H T H T T H H H H T H H H H
H H H T T H H T H T T T T H H H H H H H H H H H T
T T T T T H H T T H T T H T H T T T H T T H H T T
H H H T H T T T H H H H T H T T H T T T T T T T H
H H T H T H T T T H T T H T H T H H H T T H H H H
T T T H T T T H H T H T T T H H H H H T H H T H T
H T H H T H T H H T H H T H T T H T H H H T T H T
T H H H T H H T H T H T H T H H T T T T T T H H H

Worksheet 2

Trials by guessing

Ten people were asked to tell whether a cola was Coke or not. If they had all guessed, the result could have been anything from zero out of ten correct all the way to ten out of ten correct.

This graph shows the results of 1000 trials by pure guessing.

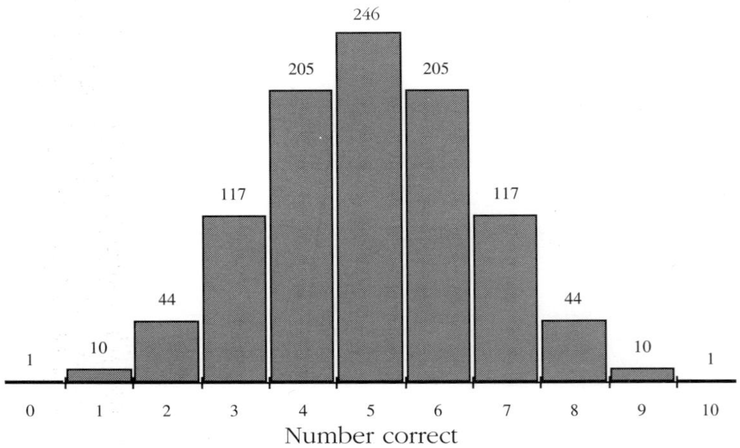

1 Complete this table to show the percentages of the total 1000 who were on or above the different numbers of correct answers out of ten.

Number correct	0	1	2	3	4	5	6	7	8	9	10
Percentage on or above	100	100	99								

2 If five out of ten people were correct, is it likely that they were all guessing? Why?

3 If nine out of ten were correct, is it likely that they were all guessing? Why?

4 If eight out of ten were correct, do you think it likely that they were all guessing? Explain.

5 How many out of ten correct would you accept as proof that they were not all guessing?

6 How likely is it that ten out of ten will be correct if all are guessing?

Investigating the spread for different sample sizes

Sample of 5

Number correct	0	1	2	3	4	5
Frequency						

Sample of 10

Number correct	0	1	2	3	4	5	6	7	8	9	10
Frequency											

Sample of 20

Number correct	0–1	2–3	4–5	6–7	8–9	10–11	12–13	14–15	16–17	18–19	20
Frequency											

Sample of 30

Number correct	0–2	3–5	6–8	9–11	12–14	15–17	18–20	21–23	24–26	27–29	30
Frequency											

Conclusion: